Technologien für die intelligente Automation

Technologies for Intelligent Automation

Band 15

Reihe herausgegeben von

inIT - Institut für industrielle Informationstechnik, Lemgo, Deutschland

Ziel der Buchreihe ist die Publikation neuer Ansätze in der Automation auf wissenschaftlichem Niveau, Themen, die heute und in Zukunft entscheidend sind, für die deutsche und internationale Industrie und Forschung. Initiativen wie Industrie 4.0, Industrial Internet oder Cyber-physical Systems machen dies deutlich. Die Anwendbarkeit und der industrielle Nutzen als durchgehendes Leitmotiv der Veröffentlichungen stehen dabei im Vordergrund. Durch diese Verankerung in der Praxis wird sowohl die Verständlichkeit als auch die Relevanz der Beiträge für die Industrie und für die angewandte Forschung gesichert. Diese Buchreihe möchte Lesern eine Orientierung für die neuen Technologien und deren Anwendungen geben und so zur erfolgreichen Umsetzung der Initiativen beitragen.

Weitere Bände in der Reihe http://www.springer.com/series/13886

Sahar Deppe

Discovery of Ill-Known Motifs in Time Series Data

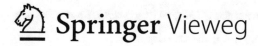

Sahar Deppe
Fraunhofer IOSB-INA
Lemgo, Germany

Dissertation, Faculty of Electrical Engineering, Computer Science, and Mathematics, University of Paderborn, 2020

ISSN 2522-8579   ISSN 2522-8587 (electronic)
Technologien für die intelligente Automation
ISBN 978-3-662-64214-6   ISBN 978-3-662-64215-3 (eBook)
https://doi.org/10.1007/978-3-662-64215-3

Responsible Editor: Alexander Gruen
This Springer Vieweg imprint is published by the registered company Springer-Verlag GmbH, DE part of Springer Nature.
The registered company address is: Heidelberger Platz 3, 14197 Berlin, Germany

Abstract

By continuous advancements in several fields of science and technology during the last decades, data mining and machine learning tasks have gained noticeable interest. The goal of these domains is to derive meaningful information from time series data by approaches such as clustering, classification, or *motif discovery*.

Motif discovery identifies frequent unknown repeated sequences in a time series and determines meaningful, new, and unknown information without any prior knowledge. Typical objections regarding motif discovery are: defining the length of motifs, large computational complexity, determining a similarity threshold, and motif identification in streaming data. Besides these difficulties, a further challenge is to determine *ill-known* motifs. Ill-known motifs are previously unknown patterns transformed by mappings such as translation, uniform scaling, reflection, squeeze, and stretch. Additionally, they may be covered with noise or have variable lengths. The analysis of state of the art reveals that only a few motif discovery algorithms are able to detect such motifs.

This dissertation provides a method which is called *ill-Known motIf discovery in Time sEries Data (KITE)*. KITE divides the input data into subsequences of equal and variable lengths and forwards them to a novel representation method in order to approximate the data without losing information. Thus, the *Analytic Complex Quad Tree Wavelet Packet Transform* (ACQTWP) is proposed to facilitate the detection of motifs that are transformed by translation, stretch, and squeeze mappings, and also motifs covered with noise. After that, to reduce the size of data and identify motifs with variable lengths and motifs altered by uniform scaling and reflection mapping, feature extraction is performed. KITE assigns six features, namely, the first four statistical moments and the maximum and minimum value of the phase of the wavelet coefficients. Finally, the similarity between all subsequences is obtained and compared with a similarity threshold to determine motifs. In contrast to other methods, KITE automatically assigns this threshold in its similarity measurement step to reduce the number of false-negatives. From all the detected motifs, KITE excludes the misleading motifs and specifies the representative ones.

Besides KITE's contribution to time series motif discovery, new avenues for the signal and image processing domain are explored and created. The proposed ACQTWP transform applies to motif discovery as well as to several signal and image processing tasks. The efficiency of KITE is demonstrated with data sets from various domains and compared with state-of-the-art algorithms, where KITE yields the best outcomes.

Kurzfassung

Durch kontinuierliche Weiterentwicklungen in diversen Bereichen der Wissenschaft und Technik während der letzten Jahrzehnte haben Data Mining und maschinelle Lernverfahren deutlich an Interesse gewonnen. Ziel dieser Domänen ist die Ableitung von aussagekräftigen Informationen aus Zeitreihendaten durch Ansätze wie Clustering, Klassifizierung oder Motif-Erkennung. Die Motif-Entdeckung identifiziert häufige, unbekannte, wiederkehrende Sequenzen in einer Zeitreihe und ermittelt ohne Vorkenntnisse sinnvolle, neue und unbekannte Informationen.

Typische Hürden der Motif-Entdeckung sind: die Definition der Länge von Motifs, große Rechenkomplexität, Bestimmung einer Ähnlichkeitsschwelle und die Motif-Identifikation in Streaming-Daten. Zusätzlich zu diesen Hindernissen besteht eine weitere Herausforderung darin, Ill-Known Motifs zu bestimmen. Ill-konwn Motifs sind bisher unbekannte Muster, die durch affine Abbildungen wie Translation, Skalierung, Reflexion, Dehnung und Stauchung transformiert werden. Darüber hinaus können sie mit Rauschen überlagert sein oder eine variable Längen haben. Eine Analyse des Stands der Technik zeigt, dass nur wenige Motif-Erkennungsalgorithmen in der Lage sind, solche Motifs zu erkennen. In dieser Dissertation wird eine Methode vorgestellt, die als "ill-known motif discovery in time series data" (KITE) bezeichnet wird. KITE unterteilt die Eingabedaten in Teilsequenzen gleicher und variabler Länge und leitet diese an ein neuartiges Repräsentationsverfahren weiter, um die Daten ohne Informationsverlust zu approximieren. So wird das Analytic Complex Quad Tree Wavelet Packet Transforms (ACQTWP) vorgeschlagen, um die Erkennung von Motifs zu erleichtern, die durch Translations-, Dehnung - und Stauchungsabbildungen transformiert werden, und auch von Motifs, die mit Rauschen bedeckt sind. Um die Datengröße zu reduzieren, Motifs mit variabler Länge und durch Skalierungsmapping veränderte Motifs zu identifizieren, wird anschließend eine Merkmalsextraktion durchgeführt. KITE extrahiert sechs Merkmale, die ersten vier statistischen Momente sowie den Maximal- und Minimalwert der Phase der Wavelet-Koeffizienten. Schließlich wird die Ähnlichkeit zwischen allen Untersequenzen ermittelt und mit einer Ähnlichkeitsschwelle verglichen, um Motifs zu identifizieren. Im Gegensatz zu anderen Methoden vergibt KITE diesen Schwellenwert automatisch um die Anzahl der Falsch-Negativen zu reduzieren. Basierend auf allen erkannten Motifs, schließt KITE irreführende Motifs aus und spezifiziert die repräsentativen Motifs.

Neben dem Beitrag von KITE zur Entdeckung von Zeitreihenmotifs werden neue Wege auf dem Gebiet der Signal- und Bildverarbeitung erforscht und geschaffen. Die vorgeschlagene ACQTWP-Transformation ist sowohl für die Motifsuche als auch für diverse Signal- und Bildverarbeitungsaufgaben anwendbar. Die Effizienz von KITE wird anhand von Datensätzen aus verschiedenen Domänen demonstriert und mit State-of-the-Art-Algorithmen verglichen, wobei KITE die besten Ergebnisse liefert.

Acknowledgement

Many hours and efforts have been invested in this work; nevertheless, it would not be possible without others' support. I wish to express my sincere appreciation to my supervisor Prof. Dr.-Ing. Volker Lohweg at the Institute Industrial IT (inIT) for many scientific discussions and cooperation resulted in several publications. Without his guides and persistent help, completing this dissertation would not have been realized. Furthermore, I wish to express my deepest gratitude to my supervisor Prof. Dr. rer. nat. Eycke Hüllermeier at Paderborn University for his support, prompt feedback and helping me to achieve the goal of this work.

I also would like to pay my special regards to Prof. Dr. rer. nat. Helene Dörksen and Natalia Moriz for valuable discussions on mathematical problems. I'm deeply indebted to Dr.-Ing. Uwe Mönks for sharing his experiences and giving valuable advice.

I'd also like to extend my gratitude to my colleagues and friends in the working group of image processing, pattern recognition, and sensor and information fusion at the research institute inIT. My special thanks go to Martyna Bator, Mark Funk, Alexander Dicks (all inIT), and Eugen Gillich at Coverno GmbH, for their friendship, cooperative work, and invaluable assistance during my study.

I also thank the International Graduate School of Intelligent Systems in Automation Technology (ISA), which is run by the Faculty of Computer Science, Electrical Engineering and Mathematics and the Faculty of Mechanical Engineering of the University of Paderborn and the Institute Industrial IT (inIT) of the OWL University of Applied Sciences and Arts (TH OWL), for founding this dissertation.

Most importantly, none of this could have happened without my parents, brother, and parents-in-law. They kept me going on and this work would not have been possible without their understanding, encouragement, and support.

Finally, I wish to thank for the support and great love of my family, my husband, Jan; my son, Theo. I am extremely grateful to their sacrifices and patience that cannot be underestimated.

Lemgo, December 2020   Sahar Deppe

Contents

Nomenclature

Chapter 1

τ Similarity threshold

$dist(.)$ Distance measure

T Time series

KITE ill-**K**nown Mot**I**f Discovery in **T**ime S**E**ries Data

Chapter 2

$^{s}d[n]$ Wavelet coefficients of the DTCWT in scale s

Im   Imaginary part of a signal

Re   Real part of a signal

$\phi_{a,b}(t)$ Scaling functions of tree A and tree B in DTCWT

$\Psi(j\omega)$ Fourier transform of $\psi(t)$

$\psi_{a,b}(t)$ Wavelet functions of tree A and tree B in DTCWT

$\mathbf{S}_K$ Set of K-frequent motif

$G[z]$ Z-transform of the filter $g[n]$

$g_{a,b}(t)$ Low-pass filters of tree A and tree B in DTCWT

$h_{a,b}(t)$ High-pass filters of tree A and tree B in DTCWT

S Shift occurred in time series $x[n]$

s Scales of the wavelet transformation

$s[n]$ A discrete subsequence

$w[n]$ Sliding window

$x[n]$ A discrete time series or signal

DTCWT Dual Tree Complex Wavelet Transform

DWT Discrete Wavelet Transform

Chapter 3

max(.) The maximum operator

min(.) The minimum operator

ARIMA Auto-Regressive Integrated Moving Average

ARMA Auto-Regressive Moving Average

CNN Convolutional Neural Networks

DFA Deterministic Finite Automaton

DFT Discrete Fourier Transform

DTW Dynamic Time Warping Distance

EDR Edit Distance on Real sequence

ED Euclidean Distance

GCT Generalised Circular Transform

LCSS Longest Common SubSequence

PAA Piecewise Aggregate Approximation

PCA Principal Component Analysis

PLA Piecewise Linear Approximation

QWT Quaternion Wavelet Transform

SAX Symbolic Aggregate approXimation

SIFT Scale-Invariant Feature Transform

SN Scattering Networks

STFT Short-Time Fourier Transform

SVD Singular Value Decomposition

US Uniform Scaling Distance

WHT Walsh Hadamard Transform

Chapter 4

DAME Disk Aware Motif Enumeration

EMMA Enumeration of Motifs through Matrix Approximation

HIME HierarchIcal based Motif Enumeration

MDL Minimum Description Length

MK Moen-Keogh algorithm

MOEN Enum. Efficient Motif Enumeration

Mr.Motif Multi resolution Motif

SOM Self-Organising Map

SWAB Sliding Window And Bottom-up method

VLMD Variable Length Motif Discovery

kBMD k Best Motif Discovery

Chapter 5

$^s\theta_Q$ Phase of the ACQTWP's approximation coefficients

$^s\widetilde{BQ}_{A,i}$ Normalised best selected approximation nodes of WPT A

$^s\widetilde{BQ}_{B,i}$ Normalised best selected approximation nodes of WPT B

$^s\widetilde{BR}_{A,i}$ Normalised best selected detail nodes of WPT A

$^s\widetilde{BR}_{B,i}$ Normalised best selected detail nodes of WPT B

$^s\widetilde{Q}_C[n]$ Normalised complex approximation coefficients of the ACQTWP transform

$[\,^aGe_i]$ Equivalence class ^aGe

$[\,^aGo_i]$ Equivalence class ^aGo

$[C_E]$ Equivalence class C_E

$[Q]$ Equivalence class of all approximation coefficients

$[R]$ Equivalence class of all detail coefficients

$[x]$ Equivalence class $[x]$

β_x Shape factor of signal $x[n]$

$\beta_{A,J}$ Shape factor of the J-th node in WPT A

$\downarrow 2^e$ Even downsampler

$\downarrow 2^o$ Odd downsampler

$\mathcal{O}(.)$ Time complexity of an algorithm -big O notation

τ_Δ Kurtosis threshold

^1m_R First-representative motif

^Km_R Kth-representative motif

$^s\Delta_{A,l}$ Minimum of the difference between the shape factor of l-th node in WPT A and signal $x[n]$

${}^{s}\Delta_{\mathrm{B},l}$ Minimum of the difference between the shape factor of the l-th node in WPT B

${}^{s}\eta(\mathbb{C}_i)$ Energy to entropy ratio of the i-th node of ACQTWP

${}^{s}\mathbb{C}_i$ Complex coefficients of the ACQTWP transform

${}^{s}BQ_{\mathbb{C}}$ Best complex approximation node in scale s

${}^{s}BR_{\mathbb{C}}$ Best complex detail node in scale s

${}^{s}C[n]$ Coefficients of WPT A

${}^{s}C^{\mathrm{I}}[n]$ Coefficients of inverse WPT A

${}^{s}D[n]$ Coefficients of WPT B

${}^{s}E_{\mathrm{c}}$ Normalised energy of the i-th node of ACQTWP

${}^{s}En(\mathbb{C}_i)$ Entropy of the i-th node of ACQTWP

${}^{s}g'_{\mathrm{a,b}}$ Synthesis low-pass filter of WPT A and WPT B

${}^{s}g_{\mathrm{a,b}}$ Low-pass filter of WPT A and WPT B

${}^{s}h'_{\mathrm{a,b}}$ Synthesis high-pass filter of WPT A and WPT B

${}^{s}h_{\mathrm{a,b}}$ High-pass filter of WPT A and WPT B

${}^{s}Q_{\mathbb{C}}$ Complex approximation coefficients of ACQTWP transform

${}^{s}Q_{\mathrm{A},J}$ Approximation coefficients of WPT A

${}^{s}R_{\mathrm{A},J}$ Detail coefficients of WPT A

${}^{s+1}\phi_{\mathrm{a},2J+(1..3)}(t)$ WPT A scaling functions

${}^{s+1}\psi_{\mathrm{a},2J+(1..3)}(t)$ WPT A wavelet functions

E_{D} Energy density of the ACQTWP's coefficients

H_{SD} Histogram of similarity degrees

J_{d} J_{d} distance

l_{d} Pre-defined motif length

l_{nd} Non pre-defined motif length

O_{d} Overlapping degree

S_{C} Set of all equivalence approximation coef. and signal $x[n]$

SD Matrix of similarity degrees

ACQTWP Analytic Complex Quad Tree Wavelet Packet Transform

BNS Best Node Selection

Chapter 6

CR Correct motif discovery rate

e Classification error by LDA

$F-M$ F-Measure

fn False negative

fp False positive

Pr Precision

Sn Sensitivity

tn True negative

tp True positive

LDA Linear Discriminant Analysis

MFPC Modified-Fuzzy-Pattern-Classifier

SNR Signal-to-noise ratio

1 Introduction

The increasing importance of technology in different aspects of human life and the accessibility to the internet result in an accumulation of a huge amount of data. According to IBM [IBM20], over 2.5 quintillion bytes of data are created every single day, obtained from different fields such as economics, medicine and epidemiology, industry and telecommunications, geographical and physical science [PfL18, ElB18, CTC⁺19, SGO20]. These data are mainly collected from e.g. the internet or various sensors in the form of time series, which is an ordered set of numbers measured at successive and regular time intervals [Fu11].

Examples of time series are the variation of stock indexes [DLC⁺19], behaviour analysis of an insect [MKZ⁺09, LaS18], yearly obtained meteorological data [PCB18], the brain activity of a patient [FWDC⁺18], and the measured volume of liquids in a bottling process [BDD⁺19]. Other types of data, such as images or videos, are not generally considered as time series due to their format. However, these sources can be converted to time series [XKW⁺07, YeK09, KWX⁺09, DuA16, BLB⁺17, Tya17]. Consequently, methods that analyse data and extract information from them, such as data mining and knowledge discovery [Gab09, MaR10], have gained considerable interest in recent decades. Data mining and knowledge discovery do not represent new methods, instead their evolution has a long history. In 1700, early methods of pattern detection in data were included in Bayes' theorem [Joy03]. In 1936, Alan Turing proposed a universal machine that could perform computations similar to computers [Tur37, Tur50]. Subsequently, various concepts, methods, and theories, such as neural networks [Iva67, Gal07, Alp10], genetic algorithms [FrB70, Mit98, Alp10], and support vector machines [CoV95, CST⁺00], were proposed to improve data mining and knowledge discovery. The term Knowledge Discovery in Databases (KDD) was first coined by Gregory Piatetsky-Shapiro in 1989, and due to these various developments, KDD became an interdisciplinary area that considered several fields like time series analysis [Fu11, ToL14].

The main goal of time series analysis is to extract meaningful information from time series data by applying data mining and machine learning tasks [Fu11, ToL15, BLB⁺17, WLM⁺20]. Examples of these tasks are clustering [Alp10, AAJ⁺19], classification [Alp10, WRH18, HGD19], query by content [Bis07], prediction [Bis07, DLC⁺19], and *motif discovery* [PKL⁺02, ToL17b]. "Motif discovery aims to identify frequent *unknown* patterns in a time series without any prior knowledge about their location, shape, or quantity" [ToL14], as depicted in Fig. 1.1.

Definition 1.1 (Motif). Given a time series T of length $N \in \mathbb{N}$, a motif of length $w \in \mathbb{N}$ consists of two subsequences $m_i = (T_i, T_{i+1}, ..., T_{i+w-1})$ and $m_j = (T_j, T_{j+1}, ..., T_{j+w-1})$ obtained from time series T, such that $|\mathrm{dist}(m_i, m_j)| \leq \tau$.

© The Author(s), under exclusive license to
Springer-Verlag GmbH, DE, part of Springer Nature 2022
S. Deppe, *Discovery of Ill-Known Motifs in Time Series Data*, Technologien
für die intelligente Automation 15, https://doi.org/10.1007/978-3-662-64215-3_1

Subsequences m_i and m_j are a sampling of $w \leq N$ contiguous positions of T such that $1 \leq i, j \leq N - w + 1$. The predefined similarity threshold τ is defined as $0 \leq \tau$ and dist(.) is a distance measure [PKL+02].

In literature, motifs have been addressed as recurring patterns, frequent trends or subsequences, shapes, and episodes [GLR17, IMD+18]. Regardless of these various designations, all these terms refer the same goal, namely the detection of frequent unknown patterns. The term motif originally comes from genetics and deoxyribonucleic acid (DNA) analysis, where a sequence motif is a reoccurring pattern or a binding site in DNA that presumed to have a biological significance [DaD07, ToL18].

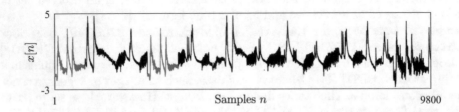

Figure 1.1: A time series gathered from a feeding analysis of a leafhopper [MKZ+09]. A pair of motifs (in green) is demonstrated with equal length.

Similarly, in the domain of time series analysis, motifs are assumed to contain valuable information about the inquiry subject [MCAER+18, ToL18]. Nowadays, motifs are parts of tasks such as rule discovery, summarisation, clustering, and classification [EsA12, ToL14, BLB+17].

In the domain of time series data mining, motifs were first introduced in 2002 [PKL+02], and since then, they have been an expanding research topic with several proposed methods and contributions. However, in spite of extensive research in time series motif discovery, several challenges and research issues remain, which serve as a motivation for this dissertation.

1.1 Motivation

A typical characteristic of time series gathered from, e.g., physical sensors is the occurrence of multiple events or unknown patterns (motifs). One of the issues regarding the detection of these motifs is that various mappings may alter their shape. As an example, motifs may be mapped by affine transformations like reflections or translations [BrM80].

Definition 1.2 (Affine transformations). An affine transformation or mapping is defined as the composition of a translation and a linear transformation. An affine mapping $f : \mathbb{R} \to \mathbb{R}$ on a time series T is represented by $f(T) = \boldsymbol{A}T + S$, where $\boldsymbol{A}$ is an invertible matrix known as transformation matrix and S is a translation vector. In particular, the following cases are obtained [Sig02]:

- $\boldsymbol{A} = \boldsymbol{I}$, where $\boldsymbol{I}$ is the identity matrix, the transformation matrix is translations.

- $\boldsymbol{A}^T\boldsymbol{A} = \boldsymbol{I}$, the transformation matrix represents the congruence transformations (e.g. translation, reflection, and rotation (2D data)) [BSM$^+$01].

- $\boldsymbol{A}^T\boldsymbol{A} = k\boldsymbol{I}$, for the ratio $k \in \mathbb{R}^+$, the transformation matrix is the similarity transformations (e.g. translation, scaling, reflection, and rotation (2D data)) [BSM$^+$01].

- $det(\boldsymbol{A}) \neq 0$, the transformation represents affine mappings (e.g. translation, reflection, stretch, squeeze, and rotation (2D data)).

Motifs affected by such transformations are designated as *ill-known* motifs.

Definition 1.3 (Ill-known motifs). Assume two subsequences $m_i = (T_i, T_{i+1}, ..., T_{i+w-1})$ and $m_j = (T_j, T_{j+1}, ..., T_{j+w-1})$ are obtained from time series T. If m_i is altered by an affine transformation (cf. Def. 1.2) such that $f(m_i) = m_j$, then subsequence m_j is the ill-known version of m_i. Given the threshold $0 \leq \tau$, if $|dist(m_i, m_j)| \leq \tau$, then m_i and m_j represent an ill-known motif.

Note: Motifs covered by undesired disturbances such as noise are also considered to be ill-known motifs.

Thus, the term ill-known motifs refers to any motifs that are affected by one or several mappings similar to the definition 1.2. Examples of ill-known motifs are illustrated in Fig. 1.2. These motifs have been enlarged, translated in time and amplitude axis, and covered with noise.

Consequently, in order to detect ill-known motifs, a motif discovery is required, invariant to most of the above-stated transformations and mappings. Despite extensive research in time series motif discovery, the existing motif discovery approaches are incapable of detecting ill-known motifs subjected to more than two types of transformations [ToL17b]. Hence, a multiple invariance motif discovery method is required to extract information from such signals. Another issue regarding motif discovery is the fact that motifs are often assumed to be of fixed length. The majority of motif discovery algorithms is tailored to a fixed static length of motifs, which must be provided in advance for such algorithms to find motifs. Consequently, these methods are only able to detect one type of equal-length motifs [PKL$^+$02, FAS$^+$06, MuC15, SiB18, ZYZ$^+$18, KCK19, LuII20]. The straightforward approach for current algorithms to discover motifs of different lengths is to iterate the same algorithm multiple times (for various lengths) [NNR12, ZaY17]. Such approaches increase the time complexity and result in a vast amount of redundant information. Thus, a motif discovery is required which can tackle the problem of variable-length motifs.

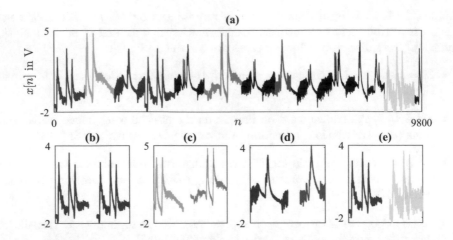

Figure 1.2: Example of ill-known patterns. (a) the signal under investigation. (b-e) motifs altered by mappings such as (b) time translation, (c) amplitude translation, (d) enlarging (squeeze/stretch), (e) noise.

Additionally, specifying diverse parameters for motif discovery algorithms constitutes another obstacle [SiR15, GLR17, GaL18, LiL19].

Most of these parameters are dependent on the tested data, and in order to provide the best set of values for these parameters, several configuration tests must be performed. Consequently, in order to tackle and overcome the described limitations, a method called *ill-Known motIf discovery in Time sEries data* (*KITE*) contributes to this work.

1.2 Goals of the Thesis

The objective of this thesis is to investigate and deal with the aforementioned limitations of the existing time series motif discovery algorithms. For this reason, the following goals can be formulated: *improving state of the art in motif discovery and filling the research gap in this domain by developing a multiple invariance algorithm that can detect ill-known motifs, as well as providing a motif discovery method that compares and discovers motifs of variable lengths. This method must be able to automatically specify the employed parameters and not be based on trial and error. The final goal is to increase the understanding of time series motif discovery and to assist the domain expert by providing the most representative motifs and excluding misleading motifs.*

1.3 Scope of the Thesis

This thesis focuses on analysing time series in the context of knowledge discovery, data mining and pattern recognition. Time series forecasting and modelling, which are part of statistical time-series analysis, are not considered. Time series analysis is mainly applied for the purpose of motif discovery for discrete time series data. In this dissertation, other motivations for analysing time series regarding pattern recognition, such as clustering, prediction, and rule discovery are not covered.

Learning algorithms employed in pattern recognition tasks, such as clustering and classification, are out of this thesis scope. Although, motif discovery also belongs to pattern recognition tasks, its procedure differs from the main steps of pattern recognition. Process of pattern recognition tasks includes the following steps: pre-processing, extraction and selection of features, and learning algorithms [Bis07, Alp10]. Pre-processing contains tasks such as normalisation, segmentation [KCH+04, LLW08], and noise reduction [Mey93, OpS89, HKV08]. Feature extraction and selection aim to derive data's distinctive aspects, reduce the amount of data, and increase performance [PNP+12]. Learning algorithms, such as classifiers or clustering algorithms, distinguish the patterns. By learning the data's characteristics from a training set, these algorithms allocate patterns of a test set to the corresponding classes or cluster the data. This learning procedure can be supervised (given the pre-defined classes, patterns are assigned to the classes using labelled data), unsupervised (the given pattern is assigned to an unknown class), and semi-supervised (the given pattern is assigned to a pre-defined class using both labelled and unlabelled data). Interested readers may consult [HoA04, BJR+15] for more information.

Nevertheless, an application of motif discovery in classification are described in Chapter 6.

The performance of the proposed approach in this dissertation is benchmarked with the algorithms within the domain of time series data mining. Other domains like biology and biomedical research are not adduced, apart from the fact that motif discovery originates from genetics. The motif detection methods [BBB+09, DTC+18] in biology and biomedical domains perform mostly on symbolic sequences and thus are not examined.

Feature extraction and selection are not the primary focus of this work. However, as feature extraction belongs to the steps of the KITE method, the features which are most suitable for this work and generalise the concept of KITE are proposed. Among the different types of features, e.g., binary and categorical, only the statistical features are regarded. Categorical features such as colour, texture, and clinical variables (temperature, blood pressure, weight) mainly depend on the application and are therefore not considered. Binary-valued features such as BRIEF [CLS+10] and BRISK [LCS11] are mostly applied in computer vision tasks [ReL15, ScG16, ZLV16] and thus are not discussed in this thesis. The feature extraction step of KITE is not limited to the proposed features, and as feature extraction is mostly application-based, several other features can be added to or omitted from this step. This matter is illustrated in Chapter 6, where adding extra features improves the motif discovery results.

It is necessary to point out that this research is not only limited to one-dimensional time-series signals and is extendible to data with higher dimensions. Applications of motif discovery for two-dimensional data are presented in this work as well.

Binary-valued signals are not evaluated in this dissertation as detecting motifs in such signals is meaningless. Finally, data streaming, which occurs in on-line applications, is not included in this dissertation as well.

1.4 Thesis' Outline

This thesis consists of seven chapters which are briefly outlined in the following:

- Chapter 1, *Introduction*, attracts the readers' attention by describing the problem at hand, the goals of this thesis, and by giving an overview of the thesis' structure.

- Chapter 2, *Preliminaries*, provides the necessary background knowledge for the main chapter (Chapter 5). An introductory concept to time series, motifs, representation, and mappings is given. Additionally, this chapter is equipped with definitions and information required to apprehend the lemmas and corollaries (mainly in Chapter 5).

- Chapter 3, *General Principles of Time Series Motif Discovery*, introduces the three main steps which are common in motif discovery and pattern recognition approaches, namely pre-processing, time series representation, and similarity measurement. Methods that belong to each step are classified and explained.

- Chapter 4, *State of the Art in Time Series Motif Discovery*, discusses several approaches and techniques related to motif discovery. The advantages and limitations of each method are explained. The state-of-the-art approaches are compared with each other with regard to several aspects, such as the definition of motif length, representation methods, the applied distance measure, and time complexity. A comprehensive description of the existing research gap is also included in this chapter.

- Chapter 5, *Distortion Invariant Motif Discovery*, is the prime chapter of this dissertation. This chapter proposes the innovative approach KITE which has five steps: pre-processing, representation and mapping, feature extraction, similarity measurement, and significant motif discovery. The core of KITE is a new shift-invariant representation method which is based on the wavelet theory, partly explained in Chapter 2. The next main step of KITE is feature extraction that results in multiple invariant (e.g. translations, scaling invariant) features. Motifs are discovered in the similarity measurement step. The threshold that is needed to quantify similarity and distinguish motifs is obtained automatically in this step. After detecting all motifs, KITE selects the significant ones by determining representative motifs and rejecting misleading motifs.

The performance of KITE regarding its computational aspect in every step is also considered in this chapter.

- Chapter 6, *Evaluation*, demonstrates the experiments that are performed in order to represent the performance of KITE. For this reason, two types of test cases, real-world and synthetic data sets, are employed. The state-of-the-art algorithms function as a benchmark for all results by considering several quality measures such as correct motif discovery rate and F-Measure.

  Two applications of KITE, anomaly detection and motif discovery on image data sets, are illustrated as well.

- Chapter 7, *Conclusion and Outlook*, concludes this dissertation and discusses the relevance of the findings. Additionally, an outlook on future perspectives is provided.

It should be noted that the peer-reviewed work of the author published in journals or conference proceedings, [ToL14, ToL15, TDDL16, ToL17b, ToL17a, ToL18], is integrated literally into this document wherever suitable.

2 Preliminaries

The definitions, mathematical notations, and tools applied throughout this thesis are given in this section. First, a general introduction to time-series is included since time series constitute this dissertation's main scope. In the context of data mining and knowledge discovery, time series analysis does not only aim to extract statistics but mainly to discover the characteristics and meaningful information from the data by employing tasks such as clustering, classification, prediction, and motif discovery [Alp10, EsA12, BLB$^+$17, ToL17b]. The core of several algorithms in these tasks depends on quantifying similarity between sequences by a distance measure [Bis07]. Thus, principles of distance and similarity measures are introduced before explaining different definitions of the time series motif. Representations and transformations belong to stated tasks' procedures either as one of the main steps or as part of pre-processing [Bis07, Alp10]. For information regarding types of representations and transformations, refer to Appendix A and [DuC04, Pou10, EsA12]. Concepts of wavelet transformations [Dau90, Mey93, BGG$^+$98, JeC01, LLZ$^+$02, Mal08, Pou10] are one of the foundations of this thesis, which is why they are explained in this chapter.

2.1 Time Series Signals

Time series or signals are gathered from various sources or sensors providing different information. Such sensors can be applied in their traditional way by employing only one sensor type (e.g., thermometer) or combining different sensors (e.g., temperature, pressure and humidity in environmental sensors). Time series are measured in continuous or discrete time.

Definition 2.1 (Continuous time series). A continuous time series is defined along a continuum of times. Thus, time series $x(t)$ where $t \in \mathbb{R}$ includes observations that are made continuously through time [Wei94].

The focus of this work is on discrete time series.

Definition 2.2 (Discrete time series). A discrete time series $x = (x_1, x_2, ..., x_N)$ is an ordered set of $N \in \mathbb{N}$ real-valued variables over discrete time points. In this work, a discrete time series is given by $x[n]$ for $1 \leq n \leq N$, where $x[1] = x_1$, $x[2] = x_2, ...,$ and $x[N] = x_N$; $x[N] = x_N$ is the most recent value [OpS89].

Next example illustrates a time series gathered from a study in the field of ethology.

© The Author(s), under exclusive license to
Springer-Verlag GmbH, DE, part of Springer Nature 2022
S. Deppe, *Discovery of Ill-Known Motifs in Time Series Data*, Technologien
für die intelligente Automation 15, https://doi.org/10.1007/978-3-662-64215-3_2

Example 1. A part of an Electrical Penetration Graph (EPG) signal that traces the feeding behaviour of a leafhopper is depicted in Fig 2.1 [StW09]. The signal is gathered by gluing a thin wire to the insect's back, completing the circuit through a host plant, and measuring fluctuations in voltage level [MKZ$^+$09, StW09].

The study conducted in [StW09] aimed to characterize the feeding behaviour of leafhoppers to analyse their harm to plants such as sugar beet, and tomato.

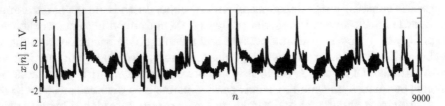

Figure 2.1: An Electrical Penetration Graph (EPG) signal gathered from a leafhop-per [MKZ$^+$09].

The terms time series and signal are applied interchangeably for the discrete type of time series or signals in this work.

Besides one-dimensional time series, there are also m-dimensional ones.

Definition 2.3 (m-dimensional time series). An m-dimensional time series $\mathbf{X}_m$ is an $m \times N$-matrix, m and $N \in \mathbb{N}$, where m is the number of rows or dimensions and N is the number of columns or the length of the time series in each dimension [Mad97, ToL18].

Such time series are common in, e.g. motion capture or trajectory tasks, where time series are stored in the xyz- or xy-coordinates.

Time series can be collected as streaming or batch data. Streaming time series are gathered from on-line applications continuously [Keo03, LLW08, LLO12, JZP$^+$19]. Batch or off-line time series $x[n]$ has a fixed length and is not infinite $n \neq \infty$. Moreover, all data points are available simultaneously.

Time series are compounded by a notable number of data, such as meteorological time series [PCB18]. In such data, it is preferable to analyse the local rather than the global properties. Therefore, subsections of time series, also referred to as subsequences are considered as well.

Definition 2.4 (Subsequence). A subsequence $s = (x_k, x_{k+1}, ..., x_{k+m-1})$ of length $m \leq N$ for all $1 \leq k \leq N - m + 1$ is a part of a time series $x[n]$ of length $N \in \mathbb{N}$. In this work, a subsequence is denoted by $s[n]$ with $1 \leq n \leq m$, where $s[1] = x_k$, $s[2] = x_{k+1}$, ..., and $s[m] = x_{k+m-1}$. Thus, a subsequence comprises contiguous time instants of a time series [PKL$^+$02, ToL15].

In general, various subsequences can be derived from time series $x[n]$. These sub-sequences are depicted by $s_i[n]$ for $i \in \mathbb{N}$ is the number of subsequences. The most common method to obtain subsequences from a time series is by using a sliding window [HCZ$^+$13, MuC15, LZP$^+$18, ToL18, ImK19, KGK19].

Definition 2.5 (Sliding window). Given a time series $x[n]$ of length $N \in \mathbb{N}$ and a window or a subsequence $w[n]$ of length $l \in \mathbb{N}$, all possible subsequences $s_i[n]$ of length l where $1 \le i \le N - l + 1$ are extracted by sliding the window $w[n]$ across $x[n]$ [KCH$^+$04].

2.2 Distance and Similarity Measure

Almost every time series data mining task requires a notion of quantifying the similarity between series. The similarity measure aims to find out to what extent two objects (time series) are alike. Similarity measures take large values when two objects are highly similar and receive either zero or negative values for dissimilar objects. In the case of distance functions, the opposite is true. Distance is a numerical description that shows to which extend objects are apart from each other, and it normally refers to a physical length [Spi15]. A distance function follows the same concept.

Definition 2.6 (Distance function). A distance function $dist$ is a function on a set A, where $dist : A \times A \to \mathbb{R}^+$ and it satisfies the three following conditions [DeD09, WFBS14]:

- Non-negativity: $dist(a, b) \ge 0, \ \forall a, b, \in A$;

- Identity: $dist(a, b) = 0 \Leftrightarrow a = b$;

- Symmetry: $dist(a, b) = dist(b, a)$.

Distance functions are called metric if they satisfy the triangle inequality [Sch90, KhK11] in addition to the above mentioned constraints.

- Triangle inequality $dist(a, c) \le dist(a, b) + dist(b, c), \ \forall a, b, c \in A$.

For metric distance functions, it is possible to arithmetically average, add, or subtract distances to compute new distances [DeD09]. This is not possible for similarities; although similarity measures are symmetric, they are not metric. Despite the differences between these two terms (distance functions and similarity measures), they are often regarded as the same in the time series data mining community. Thus, in the context of data mining, a similarity measure quantifies the similarity between two time series according to their distance [Spi15, BaR16]. Further information and various types of distance functions are explained in Sec. 3.3.

Besides similarity and distance measures, various tests (e.g. Kolmogrov-Smirnov [MJ51]) and shape factors or parameters (e.g. skewness, kurtosis [BaM88]) are available to compare the shape of the distribution of different signals.

Definition 2.7 (Similarity of signals' shape). Let R_k be an equivalence relation on $L^2(\mathbb{R})$. The relation R_k is defined by $(x_1, x_2) \in R_k$ iff $\beta_{x_1} \cong \beta_{x_2}, \ \forall x_1, x_2 \in L^2(\mathbb{R})$ and $\beta_{x_1}, \beta_{x_2} \in \mathbb{R}$. Thus, two signals x_1 and x_2 are considered similar (in shape) if their shape factor β is equivalent.

2.3 Time Series Motif

In literature, there are two classes of definitions for motifs, namely the K-frequent motif [PKL$^+$02] for $K \in \mathbb{N}$ and the nearest-neighbour motif [YKM$^+$07]. The concept of the K-frequent motif is based on the subsequence match and a user-defined positive threshold $\tau \in \mathbb{R}^+$. Based on the definition in [PKL$^+$02], K-frequent motifs are characterised by the number of matches for each subsequence. Besides a threshold $\tau \in \mathbb{R}^+$, the motif length $m \in \mathbb{N}$ must be provided to obtain such motifs.

Definition 2.8 (K-frequent motifs [PKL$^+$02]). Given a time series $x[n]$, a subsequence of length $m \in \mathbb{N}$, and a threshold τ, the K-frequent motif is the subsequence $s^K[n]$ that has the highest count of matches.

Therefore, $s^1[n]$ is the most frequently occurring subsequence in the signal $x[n]$. All of the subsequences that match, e.g. $s^1[n]$, must be mutually exclusive. Otherwise, these subsequences share most elements, which leads to very similar frequent motifs.

Definition 2.9 (Equivalence class of K-frequent motif). Equivalence class $[s^K] = \{\forall s_j[n] \subset x[n] \; : \; |dist(s^K[n], s_j[n])| \leq \tau\}$ includes all of the subsequences that match $s^K[n]$, and $0 < \mathrm{Card}([s^K]) < \mathrm{Card}([s^1])$, where $\mathrm{Card}(.)$ denotes the cardinality of a set. Additionally, $[s^1] \cap [s^2] \cap ... \cap [s^K] = \varnothing$.

A graphical representation of K-frequent motifs is given in Fig. 2.2, where two sets of motifs are detected. The equivalence class of the first-frequent motif is presented in green, while the equivalence class of the second-frequent motif is presented in red.

The alternative definition, the nearest-neighbour motif [YKM$^+$07], is based on the idea that two subsequences close in the distance should be considered a motif. Fig. 2.3 represents this type of motifs, where a pair of nearest-neighbour motifs is depicted in green.

Definition 2.10 (Nearest-neighbour motif [YKM$^+$07]). Consider a time series $x[n]$ of length $N \in \mathbb{N}$ and two subsequences $s_1[n]$ and $s_2[n]$ of length $m \in \mathbb{N}$, $m < N$, where $s_1[n], s_2[n] \subset x[n]$. Let the first elements be $s_1[i]$ and $s_2[j]$ and the last elements be $s_1[i+m-1]$ and $s_2[j+m-1]$, so for these two subsequences $i < i+m \leq j \leq N-m+1$. Subsequently, these two subsequences are nearest-neighbour motifs if $dist(s_1[n], s_2[n])$ is at its minimum among all other subsequences.

For the nearest-neighbour motifs, Def. 2.10 [YKM$^+$07], only one parameter must be defined: the length of motifs. This definition, Def. 2.10 [YKM$^+$07], does not imply the frequency of occurring motifs.

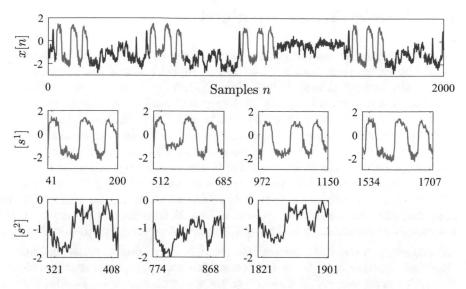

Figure 2.2: K-frequent motifs: the equivalence class of the first-frequent motif, $[s^1]$, is presented in green and the equivalence class of the second-frequent motif, $[s^2]$, is depicted by red.

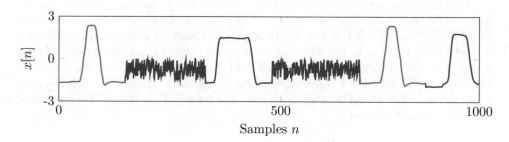

Figure 2.3: In Green: a pair of nearest-neighbour motifs with the highest similarity to each other.

Each of the motif definitions above has its advantages and disadvantages. A top 5-frequent motif can be a noise subsequence where the nearest-neighbour motif can represent a pattern that occurred only once. Thus, none of these definitions claims to be better than the other.

In this work, the discovery of motifs is based on the combination of both introduced definitions as given in Chapter 5.

2.4 Wavelet Transformations

As given in [Ada10], "wavelet theory is one of the most modern theories developed mostly by researchers, such as Yves Meyer, Ingrid Daubechies, Stéphane Mallat, and Albert Cohen" [Mal89, Dau90, Mey93, CoR95]. A wavelet is a mathematical function, a waveform with a time-limited extent and an average of zero, that describes the time-frequency plane [Dau90, Mey93, BGG$^+$98, JeC01, LLZ$^+$02, Mal08, Pou10, Ada10]. A family of wavelets can be constructed from a "mother wavelet" in order to extract time and frequency information. The wavelet members are formed by changing the values of a translation factor and by expanding a scale parameter. For general information about wavelet transformations, refer to Appendix A, Chapter 8. It is possible to stretch, compress, and translate the mother wavelet by assigning different values to its scale as well as its translation parameters to obtain the required resolution for capturing the signal characteristics [BGG$^+$98, Ada10].

Consequently, wavelet transformations assist in determining motifs that have been stretched or squeezed. They also constitute an appropriate mathematical tool in case of analysing non-stationary or transient phenomena. Besides the stated aspects, other properties such as a flexible time-frequency resolution and a perfect reconstruction make them a proper tool for this work.

The most commonly-applied form of wavelet transformations is the Discrete Wavelet Transform (DWT) [Mey93]. However, the DWT is shift-variant due to the downsampling operations [Mey93]. This means that small time-shifts in the input signal cause changes in DWT coefficients. This issue of shift invariance is also addressed by Kingsbury in [Kin01]. He proposed the Dual Tree Complex Wavelet Transform (DTCWT) [Kin01] to overcome this problem. However, due to the definition in [SBK05], the DTCWT is not completely shift-invariant [ToL15]. The DTCWT is only shift-invariant in the first scale and scales greater than one are not able to accomplish this property. The DTCWT and its problem regarding time translations are explained in the following section. It should be noted that the lemmas, proofs, and explanations published by the author in [ToL15, ToL18] are integrated literally into this section.

2.4.1 Dual Tree Complex Wavelet Transform (DTCWT)

The DTCWT [Kin01, SBK05] is proposed to overcome the shortcomings of the DWT, such as being shift variant and the lack of directionality for 2D DWT. The structure of the DTCWT is based on two parallel filter bank trees: tree A and tree B. To have a complex representation, tree A represents the real part and tree B provides the imaginary part [SBK05]. Fig. 2.4 represents DTCWT's structure.

Definition 2.11 (DTCWT's wavelet and scaling functions). Let $\psi_a(t)$, $\psi_b(t)$, and $\phi_a(t)$, $\phi_b(t)$ be the wavelet and scaling functions of the DTCWT. The wavelet and scaling functions in tree A, $\forall n \in \mathbb{N}$ are given by [BGG^{+}98, SBK05]

$$\psi_{\mathrm{a}}(t) = \sqrt{2} \sum_{n=0}^{M} h_{\mathrm{a}}[n]\phi_{\mathrm{a}}(2t - n), \qquad \phi_{\mathrm{a}}(t) = \sqrt{2} \sum_{n=0}^{M} g_{\mathrm{a}}[n]\phi_{\mathrm{a}}(2t - n),$$

where filter h_{a} and g_{a} are high- and low-pass filters and $s \in \mathbb{N}$ is the number of scales. All filters are causal, so $h_{\mathrm{a,b}}[n] = 0$ and $g_{\mathrm{a,b}}[n] = 0$ for $n < 0$. Similarly, for tree B, the wavelet and scaling functions are defined [ToL18].

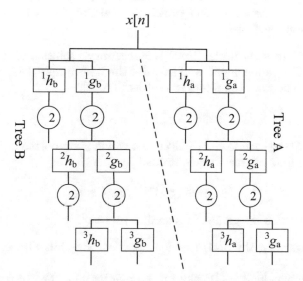

Figure 2.4: Filter bank structure of the DTCWT [SBK05]. Low- and high-pass filters of tree A and B are given by $^{s}g_{\mathrm{a}}$, $^{s}h_{\mathrm{a}}$ and $^{s}g_{\mathrm{b}}$, $^{s}h_{\mathrm{b}}$, respectively. $\downarrow 2$ depicts the downsampling operator.

Due to their design [SBK05], the filters of the DTCWT form a Hilbert pair [Hil53, Bed62].

Definition 2.12 (Hilbert pair). Two wavelets ψ_{a} and ψ_{b} with the following property [Hil53, Bed62, Sel01]

$$\Psi_{\mathrm{a}}(j\omega) = \begin{cases} -j\Psi_{\mathrm{b}}(j\omega), & \omega > 0, \\ j\Psi_{\mathrm{b}}(j\omega), & \omega < 0, \end{cases}$$

are called the Hilbert pair, where $\Psi(j\omega)$ is the Fourier transform of $\psi(t)$.

Theorem 1 (Half-sample delay [Sel01]). Wavelets ψ_{a} and ψ_{b} form a Hilbert pair if the filters $^{s}g_{\mathrm{a}}$ and $^{s}g_{\mathrm{b}}$ satisfy the condition [Sel01],

$$^{s}G_{\mathrm{a}}(e^{j\omega}) = {^{s}G_{\mathrm{b}}}(e^{j\omega})e^{-j\frac{\omega}{2}}. \tag{2.1}$$

Eq. 2.1 represents the half-sample delay condition for two low-pass filters ${}^{s}g_{\mathrm{a}}$ and ${}^{s}g_{\mathrm{b}}$. Another representation of Eq. 2.1 is in the form of magnitude and phase functions, given by [AbS04, SBK05]:

$$|{}^{s}G_{\mathrm{a}}(e^{j\omega})| = |{}^{s}G_{\mathrm{b}}(e^{j\omega})|, \qquad \angle {}^{s}G_{\mathrm{a}}(e^{j\omega}) = \angle {}^{s}G_{\mathrm{b}}(e^{j\omega}) - \frac{1}{2}\omega, \qquad (2.2)$$

Proof. The proof is presented by Selesnick [Sel01].   □

This means that the scaling low-pass filters must be offset from another by half a sample [YuO05]. Moreover, Yu and Ozkaramanli [Sel01, Kin03, YuO05] have proven that "Eq. 2.2 is the necessary and sufficient condition for two wavelets to form a Hilbert transform pair" [ToL15].

Definition 2.13 (q-shift filters [Kin03]). Kingsbury's solution of the filters' design for the DTCWT is called q-shift and satisfies the half-sample delay condition given in Theorem 1. The low-pass filters are set as [Kin03]

$$ {}^{s}g_{\mathrm{a}}[n] \; = \; {}^{s}g_{\mathrm{b}}[M-1-n]. \qquad (2.3)$$

Here, $M \in \mathbb{N}^{+}$ is the even length of filter ${}^{s}g_{\mathrm{b}}$, which is supported on $0 \leq n \leq M-1$. The high-pass filters of tree A are obtained by

$$ {}^{s}h_{\mathrm{a}}[n] = (-1)^{1-n}\; {}^{s}g_{\mathrm{a}}[1-n]. \qquad (2.4)$$

For tree B, the filters are gained in a similar manner.

Information regarding the design and properties of q-shift filters is given in Appendix 9.2.

Based on the noble identities [Vai90] (cf. Appendix 9.1), the low-pass filters of tree A can be presented by [SBK05]

$$ {}^{s}G_{\mathrm{a,T}}[z] \; = \; {}^{1}G_{\mathrm{a}}[z]\, {}^{2}G_{\mathrm{a}}[z^{2}]\, {}^{3}G_{\mathrm{a}}[z^{3}]... \,{}^{s}G_{\mathrm{a}}[z^{2^{s-1}}], $$

where $G[z] = \sum\limits_{n=0}^{\infty} g[n]z^{-n}$ and the index T stands for the total filters. The low-pass filters of tree B ${}^{s}G_{\mathrm{b,T}}[z]$ are obtained in a similar manner. In order to achieve the half-sample delay theorem at each scale, "the filters of tree A translated by 2^{s} must fall midway between the translated filters of tree B" [SBK05]. For the first scale, the translated filters of ${}^{1}G_{\mathrm{a,T}}[z]$ by two are required to fall midway between the translated filters of ${}^{1}G_{\mathrm{b,T}}[z]$ by two, so [SBK05]

$$ {}^{1}G_{\mathrm{a,T}}[z] \; = \; {}^{1}G_{\mathrm{b,T}}[z]z^{-1}. \qquad (2.5)$$

For the second scale, this is obtained by [SBK05]

$$ {}^{2}G_{\mathrm{a,T}}[z] \; = \; {}^{2}G_{\mathrm{b,T}}[z]z^{-2}. \qquad (2.6)$$

For the third scale, it is ${}^{3}G_{\mathrm{a,T}}[z] = {}^{3}G_{\mathrm{b,T}}[z]z^{-4}$, and so forth. Only in Eq. 2.5, it is $^1G_{a,T}[z] = {}^1g_a[n]$ [SBK05], thus at the first scale

$$^1g_a[n] = {}^1g_b[n-1].\qquad(2.7)$$

Scales greater than one correspond to the half-sample delay theorem, e.g. [SBK05]

$$^2G_{a,T}[z] = {}^2G_{b,T}[z]z^{-2},$$
$$^1G_a[z]\,{}^2G_a[z^2] = {}^1G_b[z]\,{}^2G_b[z^2]z^{-2}.$$

Based on Eq. 2.7, $^1G_a[z] = {}^1G_b[z]z^{-1}$. Thus, $^2G_a[z^2] = {}^2G_b[z^2]z^{-1}$, which is equivalent to

$$^2G_a[z] = {}^2G_b[z]z^{-\frac{1}{2}},$$
$$^2g_a[n] = {}^2g_b[n-\frac{1}{2}].$$

Therefore, the DTCWT needs another type of filter in the first scale [SBK05]. The filters ^1g_b and ^1g_a must have one sample delay difference (cf. Eq. 2.7). Thus, as explained in [Kin01, SBK05], the same set of filters can be applied for both DTCWT's trees in the first scale, whereby one set of filters must be translated by one sample concerning the other set.

Definition 2.14 (First scale DTCWT's filter). The filters for the first scale of the DTCWT are obtained by [AbS04, SBK05]

$$^1g_a[n] = {}^1g_b[n-1].\qquad(2.8)$$

Similarly, $M \in \mathbb{N}^+$ is the even length of filter ^1g_b, which is supported on $\forall n \in \mathbb{N}$, $0 \le n \le M-1$. The first scale high-pass filters are given by [SBK05]

$$^1h_a[n] = {}^1h_b[n-1].\qquad(2.9)$$

The coefficients of the DTCWT are defined as follows.

Definition 2.15 (DTCWT coefficients). The DTCWT's wavelet and scaling coefficients of scale s, $^sd[n]$, $^sc[n]$ $n, s \in \mathbb{N}$ are obtained from both tree A and tree B, and are denoted by [SBK05]

$$^sd[n] = \mathrm{Re}(^sd[n]) + j\,\mathrm{Im}(^sd[n]),\qquad ^sc[n] = \mathrm{Re}(^sc[n]) + j\,\mathrm{Im}(^sc[n]),$$

where $d[n]$ is computed by convolving $x[n]$ with the filter $^sh_{a,b}[n]$ and then downsampling the result by factor 2^s, ($\downarrow 2^s$):

$$^sd[n] = ((x * {}^sh_a)[n]) \downarrow 2^s + j((x * {}^sh_b)[n]) \downarrow 2^s$$
$$= \sum_{k=0}^{L+M-1} x[2n-k]\,{}^sh_a[k] + j \sum_{k=0}^{L+M-1} x[2n-k]\,{}^sh_b[k],\ k \in \mathbb{Z},\qquad(2.10)$$

the filters $^sh_{a,b}[k] = 0$ for $k < 0$ are high-pass causal filters of tree A, B. $^sc[n]$ is computed by a similar procedure [ToL15].

2.4.1.1 Limitations and Deficiencies

The DTCWT has certain benefits, such as providing an analytic representation
of the signal and an efficient implementation [Kin03, SBK05]. However, it has
shortcomings, like not having a complete decomposition and being shift variant, as
explained in this section.

Deficient Decomposition

The spectrum of each DTCWT wavelet tree in the first scale is divided into two
equal parts of low- and high-pass bands. In the second scale, only the low-pass
band is decomposed into two lower equal parts: another low- and high-pass band.
This procedure continues for each wavelet tree and the rest of the scales [Kin01],
resulting in the frequency bands as depicted in Fig. 2.5. As illustrated in Fig.
2.5, the DTCWT decomposes the right side of the frequency spectrum (the low-
frequency band) in each scale, and this frequency decomposition my become a limitation for analysing some signals [BaS08, WBK09, ToL15]. As an example, signals
composed of chirps [Fla01] such as audio or biomedical signals (electroencephalog-
raphy (EEG) or electromyography (EMG)) cannot be completely analysed by the
DTCWT [TDDL16, WRH18]. If the dominant frequencies are located in the low-
frequency bands, this information will be analysed by the DTCWT.

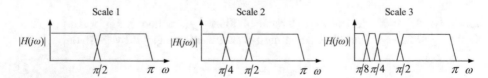

Figure 2.5: Frequency spectrum of the DTCWT's tree A in three scales. In each
scale, only the low frequency part is further decomposed.

However, if the dominant frequencies are located on the left part of the frequency
spectrum in Fig. 2.5, this information will only be analysed in the first scale.
Therefore, an alternative structure is needed in order to adapt to signals by a finer
and more adjustable frequency resolution.

Shift Invariance

Definition 2.16 (Shift or Translation). Let $f(n_1)$ be a function or a signal. The
time-shifting or translation occurs by changing the original starting point of $f(n_1)$
such that $f(n_2)$, where $n_2 = n_1 + S$ for the shift factor $S \in \mathbb{Z}$ [Ber09].

Definition 2.17 (Shift-invariant transformation). Transformation $T : X \to Y$,
$T(x[n]) = y[n]$ for the input signal $x[n]$ and the output signal $y[n]$ is called shift

invariant if it satisfies [Woo96, BRL04, Ber09]

$$T(g_s x[n]) \cong T(x[n]), \Rightarrow y[n+s] \cong y[n],$$

where the translation operator $g \in \mathcal{G}$ is given by $g_s(x[n]) = x[n+s]$, and $s \in \mathbb{Z}$ is an arbitrary shift.

The relation between the input and output must not change if a time shift (translation) is applied. Thus, the output of the translated signal must look exactly the same as the original signal. If $x[n]$ and $x[n-S]$ are the original and translated signals, shifted by $S \in \mathbb{Z}$, then for a shift-invariant wavelet transformation, the following must hold: $x[n] \cong x[n-S], \Rightarrow {}^s d[n] \cong {}^s d[n-S]$, where ${}^s d[n]$ and ${}^s d[n,S]$ are the corresponding wavelet coefficients [ToL18].

Based on Def. 2.17, the DTCWT is only shift invariant in the first scale and not in scales greater than one. The effect of time translations in the wavelet coefficients of the DTCWT is explained in the following.

Lemma 1. Assume $x[n]$ is a discrete signal and let $S_{e/o} \in \mathbb{Z}$ be shifts that occurred on signal $x[n]$, $S_{e/o}$ is even, $S_e = 2m$, or odd, $S_o = 2m - 1$, $m \in \mathbb{Z}$. The wavelet coefficients of $x[n - S_e]$ and $x[n - S_o]$ in the first scale are depicted by ${}^1 d_e[n, S_e]$ and ${}^1 d_o[n, S_o]$, respectively, and given by:

$$
{}^1 d_e[n, S_e] = \mathrm{Re}({}^1 d_e[n, S_e]) + j \, \mathrm{Im}({}^1 d_e[n, S_e])
$$
$$
= \sum_{k=0}^{L+M-1} x[2n - k - S_e] \, {}^1 h_a[k] + j \sum_{k=0}^{L+M-1} x[2n - k - S_e] \, {}^1 h_b[k],
$$

$$
{}^1 d_o[n, S_o] = \mathrm{Re}({}^1 d_o[n, S_o]) + j \, \mathrm{Im}({}^1 d_o[n, S_o])
$$
$$
= \sum_{k=0}^{L+M-1} x[2n - k - S_o] \, {}^1 h_a[k] + j \sum_{k=0}^{L+M-1} x[2n - k - S_o] \, {}^1 h_b[k].
$$

Then, the following equations hold:

$$
\forall\, x[n - S_e], \quad \begin{cases} \mathrm{Re}({}^1 d[n - \frac{S_e}{2}]) = \mathrm{Re}({}^1 d_e[n, S_e]), \\ \mathrm{Im}({}^1 d[n - \frac{S_e}{2}]) = \mathrm{Im}({}^1 d_e[n, S_e]). \end{cases}
$$

$$(2.11)$$

$$
\forall\, x[n - S_o], \quad \begin{cases} \mathrm{Re}({}^1 d[n - \lfloor \frac{S_o}{2} \rfloor]) = \mathrm{Im}({}^1 d_o[n, S_o]), \\ \mathrm{Im}({}^1 d[n - (\lfloor \frac{S_o}{2} \rfloor) + 1]) = \mathrm{Re}({}^1 d_o[n, S_o]). \end{cases}
$$

Proof. Proof is provided in Appendix 9.3. □

Lemma 2. Assume that $x[n]$ is a discrete signal and let $S_{e/o} \in \mathbb{Z}$ be shifts that occurred on signal $x[n]$, where S_e and S_o are even and odd shifts. Then for $s > 1$

$$\forall x[n - S_e], \quad \begin{cases} \text{If } mod(S_e, 2^s) = 0, & \begin{cases} \text{Re}(\ ^s d[n - \frac{S_e}{2^s}]) = \text{Re}(\ ^s d_e[n, S_e]), \\ \text{Im}(\ ^s d[n - \frac{S_e}{2^s}]) = \text{Im}(\ ^s d_e[n, S_e]). \end{cases} \\ \\ \text{Else} & \quad\quad ^s d[n] \not\cong\ ^s d_e[n, S_e]. \end{cases}$$

$$\forall x[n - S_o], \quad\quad\quad\quad\quad\quad\quad ^s d[n] \not\cong\ ^s d_o[n, S_o].$$

$$(2.12)$$

Proof. Proof is given in Appendix 9.4.   □

In [SBK05], the authors claim that the DTCWT is shift invariant, although their definition is based on the reconstructed wavelet coefficients. Accordingly, the reconstructed wavelet coefficients of the shifted signal are equivalent to the original signal. In other words, $x[n] \cong x[n - S]$, $\Rightarrow\ ^s d[n] \cong\ ^s d'[n - S]$, where $x[n - S]$ is the shifted version of $x[n]$ and $^s d'[n - S]$ is its reconstructed wavelet coefficients.

3 General Principles of Time Series Motif Discovery

The general approach towards motif discovery is depicted in Fig. 3.1. The first step, pre-processing, contains tasks such as normalisation or segmentation. The data representation step aims to reduce the data's size or to approximate it without losing relevant information. In the final step, the similarity measurement step, motifs are detected by quantifying their similarity based on a given threshold [EsA12, Mue14, ToL17b].

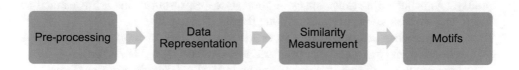

Figure 3.1: General approach of time series motif discovery comprising three steps: pre-processing, data representation, and similarity measurement.

Although Fig. 3.1 illustrates a typical motif discovery approach, not all of these methods follow the complete procedure. As an example, there are methods which only include the pre-processing and similarity measurement steps. The steps in Fig. 3.1 are described in the following sections.

3.1 Time Series Pre-Processing

Some of the common problems regarding time series in general are the size of the data, being subject to noise and outliers, as well as offset differences between time series. In motif discovery, these challenges are handled in the pre-processing step by data reduction, noise decrement, and normalisation. Data reduction aims to eliminate irrelevant parts of the data while preserving relevant information. Examples of this type of approach are sampling methods and Principal Component Analysis (PCA) (original paper in [Pea01]) [Alp10]. Noise decrement is mostly handled by methods such as digital filters or wavelet thresholding [OpS89, BGG+98, Mal08]. Amplitude or offset differences are eliminated by normalising (z-normalisation). Time series segmentation is another task that is mostly considered in the pre-processing step of several motif discovery approaches [GLR17, ZaY17, IMD+18, YWZ+19].

© The Author(s), under exclusive license to
Springer-Verlag GmbH, DE, part of Springer Nature 2022
S. Deppe, *Discovery of Ill-Known Motifs in Time Series Data*, Technologien
für die intelligente Automation 15, https://doi.org/10.1007/978-3-662-64215-3_3

3.2 Time Series Representation

A crucial problem regarding data gained from various sources such as large-scale
sensor networks, social networks, or medical and healthcare records is their massive
size [MaW15, CaL17, ElB18], which impedes the process of analysing, obtaining, or
uncovering hidden information and significant aspects of the data. As a result, efforts have been made in order to tackle this problem by performing a representation
method beforehand on the raw data.

A representation method (Def. 8.6) approximates the data accurately without
losing their main characteristics [Fu11, EsA12]. Therefore, the quality of the representation method that approximates the data is vital. A representation method
must preserve both local and global characteristics of the data. It should have a
desirable computational efficiency to handle even large data sets. Insensitivity to
noise is another demanding property, while the ability to reconstruct the original
signal after applying a representation method constitutes a further advantageous
quality. Another important aspect of time series representations is invariance to
mappings, such as affine transformations [Ber09] (Def. 8.5).

Various approaches are introduced to tackle the problem of representing data. A
comprehensive review of these approaches is included in [EsA12, Wil17].

A famous example of representation methods is the Discrete Fourier Transform
(DFT) (in French: [Fou22]), which is employed in several domains [LLR+17, SeR17].
However, DFT lacks the critical feature of time localisation. This limitation is
partly overcome by the Short-Time Fourier Transform (STFT) [BrB74, OpS89].
The main idea behind STFT is to divide a signal into blocks and to compute each
block's spectrum. This means multiplying the input signal by a fixed-length window
function and computing the Fourier transform. Apart from its straightforward
approach, the STFT has a problem defining windows to achieve a good time and
frequency resolution trade-off [Keh11].

Wavelet transformations [Mal89, Dau90, Mey93, CoR95] tackle the drawbacks of the
DFT and the STFT by scaled and shifted versions of a mother wavelet function.
The most well-known wavelet transformation is the Discrete Wavelet Transform
(DWT) [Mey93, Mal08], which has properties such as multi-resolution analysis,
perfect reconstruction, and linear computational complexity. General information
regarding wavelet transforms is provided in Appendix A. Besides the DWT, several
other types of wavelet transformations are applied in literature as a tool for time
series representations [BeK18, SMY+19].

Piecewise Aggregate Approximation (PAA), which is proposed by Keogh et al. in
[KCP+01], transforms a time series $x[n]$ of the length $N \in \mathbb{N}$ into $\widehat{x}[n]$ of length $M <
N$. Applying a sliding window of the length $w \in \mathbb{N}$, PAA divides the time series
$x[n]$ into M equal-length segments. Subsequently, each segment is represented by
its empirical mean value. PAA's drawbacks are its noise sensitivity and it's prior
definition of each section's length, which is mostly defined by several trial and
errors. The risk of missing some patterns due to the mean value representation
constitutes another weakness of this method.

A symbolic representation method called Symbolic Aggregate approXimation (SAX), which is introduced in [LKL$^+$03], converts a time series into a sequence of symbols. In order to speed up SAX, indexable Symbolic Aggregate Approximation (iSAX) is proposed in [ShK08]. SAX is not the only symbolic representation method, although this method is one of the most commonly-applied representation methods in the domain of time series motif discovery [Fu11,LLC$^+$15,ZaY17,GLR17,YWZ$^+$19]. Schäer and Högqvist [ScH12] proposed the symbolic Fourier approximation (SFA), which maps the signals into a sequence of symbols, named SFA word. SFA comprises two steps: approximation and quantisation. The approximation step represents a time series of length $N \in \mathbb{N}$ by a transformed signal of reduced length $l \in \mathbb{N}$ applying the DFT. In the quantisation step, the frequency domain is divided into frequency bins, and each Fourier coefficient is mapped to its bin. All of the bins have equal width, which leads to this approach's main problem: the inability to capture the time localisation by the DFT. This representation is mainly applied in classification tasks [Sch16].

Model-based methods are another type of representation approach. In these methods, time-series are illustrated by the relative parameters of the model that describes the data. Thus, two time series are similar if they have been generated by the same model. Examples of these models are Auto-Regressive Moving Average (ARMA) models [original article published by [Whi51]], Markov Chains [KS$^+$83] (introduced by A. Markov (1865-1922)), or Hidden Markov Models (HMM) [BaP66]. Other extensions of HMMs are presented in various literature, such as the hidden semi-Markov model [Yu16], which captures the time spent in a state. The state-based behaviour of a discrete-event system can be represented as a Deterministic Finite Automaton (DFA). Adding the event timing results in a Timed Automaton (TA) [AlD94]. Extending a TA by modelling continuous variables leads to a continuous hybrid automaton. The HyBUTLA algorithm [NSV$^+$12] is an example of such a method that automatically identifies behaviour models of its components [NSV$^+$12]. A discrete form of HyBUTLA is called BUTLA, which is suitable for off-line applications [MVN$^+$11]. OTALA [Mai14] is an extension of BUTLA that learns a TA in an on-line manner. Such approaches are mostly applied in the field of time series prediction and modelling. extbfHMMHidden Markov Models

3.2.1 Invariant Transformations

A critical challenge for most representation methods while emphasising the data's main characteristics is to be invariant under different (affine) transformations and handle distortions such as noise in the data. As described in Sec. 3.1 and in [BKT$^+$14], problems of noise, outlier, or amplitude and offset variances can be solved and removed in pre-processing. Nevertheless, other issues such as time-shifting, scaling or proportion variances, and reflection remain and must be handled differently. There are different representation methods that are invariant to affine transformations, such as translation and uniform scaling [ReB69, BrM80, Mey93, LoM02, CCB04, Mal12, HH19]. The subsequent sections provide a brief review of these methods.

Translation-Invariant Transformations

A straightforward approach to handling time translation is to align time series in advance in order to compensate for the possible shift between them. Thus, either the amount of shift is required or several trial and errors must be performed to align the time series. Another possibility to solve this issue is to apply a shift-invariant transformation.

The definition of a shift-invariant transformation is given in Def. 2.17. According to Def. 2.17, if the input signal $x[n]$ with the output $y[n]$ under the transformation T, takes some time shift so that $\breve{x}[n] = x[n+s]$ ($x[n+s]$ looks exactly the same as $\breve{x}[n]$), then it is expected that the output $y[n+s]$ is also equivalent to $y[n]$ [Yu12]. Thus, for transformation T, the input and output relation must not be changed if the input signal is translated.

One commonly-applied shift-invariant transformation is the DFT [Fou22], employed in researches such as [Sch16, ZLG17, SLK18]. It is well known that the magnitude of the Fourier transform is shift invariant [SBK05, Pou10, Ada10].

Rapid or R-transform [ReB69] is also a shift-invariant transformation that only needs two functions, f_1 and f_2, with the following operations: additions, subtractions, and absolute value [BaS10, LLG$^+$14]. A class of shift-invariant transforms that applies two functions f_1 and f_2 but with min(.) and max(.) operators is the CT-transform (certain transform) [BuM80]. A subclass of the CT-transform is a transformation where f_1 and f_2 are defined by $a+b$ and $(a-b)^2$ [MüM11, LiD12, Mül13]. The power spectrum of the Walsh-Hadamard transform [Kun79, AhR12] is also invariant to shifts. The Walsh-Hadamard's basic functions are not sinusoids, like it is the case in the DFT, but rather rectangular waveforms [ElR83]. This transformation is applied in image and signal-processing applications, such as in [SrS18, SKA18].

The Generalised Circular Transform (GCT) [LoM02] is another non-linear location-invariant transform applied in different image processing applications [LoM02, LDM04, MBM09]. This method is based on an amplitude spectrum that operates with sums of ordered period groups. This may appear to be similar to the power spectrum of the Walsh-Hadamard Transform (WHT); however, the GCT uses an absolute value spectrum. The GCT is useful in the case of periodic signals.

Wavelet transformations [Mal89, Dau90, Mey93, CoR95] are powerful representation methods, although not all of them have the shift-invariant property. Several shift-invariant wavelet transformations are introduced in various literature, such as the stationary wavelet transform [Mey93, BGG$^+$98], the shift-invariant wavelet packet [GGS$^+$09], and the Quaternion Wavelet Transform (QWT) [CCB04]. The QWT [CCB04] is based on the quaternion Fourier transform and has a magnitude as well as three phases. The first two phases of the QWT describe the shifts of the image features in the vertical and horizontal directions, while the third phase represents the texture information of the image [UmY18, GBZ18].

Mallat introduced the Scattering Networks (SN) [Mal12] for the signal classification task. Scattering networks scatter the signal information along multiple paths, with a cascade of wavelet operators implemented in a deep convolutional network

[Mal12]. These networks provide a translation-invariant representation [SiK17, OZH$^+$19].

Besides the stated methods, SAX [LKL$^+$03], PAA [KCP$^+$01], Piecewise Linear Approximation (PLA), and Singular Value Decomposition (SVD) [KJF97] are also shift invariant.

Scaling-Invariant Transformations

In order to overcome the problem of uniform scaling or proportion invariance, different methods and approaches are introduced in various literature.

Definition 3.1 (Proportion invariant (Uniform scaling invariant)). Transformation $T : X \to Y$, $T(x[n]) = y[n]$ for the input signal $x[n]$ and the output signal $y[n]$ is called proportion invariant if it satisfies

$$T(g_b x[n]) \cong T(x[n]), \ \Rightarrow y[bn] \cong y[n],$$

where the scaling operator $g \in \mathcal{G}$ is given by $g_b(x[n]) = x[bn]$, and $b \in \mathbb{Z}$, $b \neq 0$ is an arbitrary scaling factor [BrM80].

Note: In this work, the term proportion invariant is used interchanged with the term scaling invariant since the term scale is mostly applied in the context of wavelet transforms.

Commonly-applied time series representation methods in the domain of motif discovery, such as PAA, SAX, and SVD, are not proportion invariant. Nevertheless, a proportion-invariant SVD is proposed in [Uhl18].

Most of the proportion-invariant methods are utilised in image processing applications [KuS12, NRG13, PBD17]. The Mellin transform [HaL16] provides a proportion-invariant matching since its magnitude is proportion invariant. In order to provide translation and proportion-invariant transformations, the Fourier transform is combined with the Mellin transform in [FNR16, GCK$^+$18].

The log-polar transform [Pou10] is exerted in [ElE17, SaB18] since proportion and rotation are manifested in this transformation. The log-polar transform converts the magnitude spectrum of an image from the Cartesian space to the log-polar space, where rotation and proportion changes are determined by a vertical and horizontal shift. One disadvantage of this transformation is its sensitivity to the proportion factor, as large proportion factors would alter the frequency content for a recognition task.

Graph matching [YCZ$^+$15] is mostly applied to image processing applications and optimises the point-to-point correspondence. The methods based on graph matching are usually slow when the size of the data increases.

Scale-Invariant Feature Transform (SIFT) [Low99] is another type that is mainly applied in classification tasks [OZH$^+$19].

The use of Convolutional Neural Networks (CNNs) [Alp10, BuL17] with scale- and shift-invariant features is rapidly increasing [YTL$^+$16, VaP17, SiD18]. Such methods can tackle the problem of invariants, although several parameters must be defined in advance. Additionally, over-fitting, computational complexity, and the training data's size may become a problem for CNNs. Other methods that employ scale-invariant features are the bag of features and local-feature-based mesh [PBD17, HGD19].

A survey of the proportion and affine-invariant approaches is provided in [PaS17, HH19].

Scaling is classified into uniform and non-uniform scaling. In the first type, each basis vector is scaled by the same value, whereas in non-uniform scaling, each basis can either get a different scale or none. Non-uniform scaling changes the shape of the object (time series) so that the time series is stretched or shrunk [BrM80, Wol88]. Besides wavelet transformations, most of the stated transformations are not invariant to non-uniform scaling. This constitutes another reason for employing wavelet transformations as a tool in this work.

3.3 Time Series Distance and Similarity Measures

Another foremost step in motif discovery approaches is the similarity measurement. Similarity and distance measures were defined and their differences were revealed in Sec. 2.2. These measures are broadly classified into shape-based, edit-based, and model-based groups [EsA12, WMD$^+$13, WFBS14, LFW$^+$19].

Shape-based measures are the most applied measures that compare time series' total shape [BHW$^+$04, FoH10, BLB$^+$17, MCAER$^+$18]. An important class of such measures is metric measures, defined in Sec. 2.2, such as the family of L_p distances or Minkowski distances. The family of L_p distances measures the similarity between two signals based on the point-to-point concept meaning comparing between the i–th data point of one signal and the corresponding i–th point of the other signal [WMD$^+$13]. Members of the L_p distances, especially the Euclidean Distance (ED) [DeD09], are the most widely-applied distance measures in several time series data mining tasks [LZP$^+$18, GaL18, FeN19]. Another member of L_p distances is the Canberra distance [LaW66] which is a weighted version of Manhattan distance [DeD09, Cra10].

There are two main problems regarding L_p distances: these measures are not robust against time translations and scaling, even if the time series' shape is similar [DTS$^+$08], and they are sensitive to outliers and noise.

Elastic measures [EsA12, WMD$^+$13, LiB15] allow flexible comparison (one to many) between two signals' data points by quantifying the similarity between several parts of time series considering localised misalignments through some elastic adjustment. Thus, they are more flexible than the family of L_p norms or Minkowski distances. Dynamic Time Warping (DTW) [BeC94], Uniform Scaling (US) [Keo03], and the Spatial Assembling Distance (SpADe) [CNO$^+$07] belong to the elastic measures.

DTW [BeC94] is the most famous elastic measure which has been applied effectively in time series data mining tasks [SiB16, DSP$^+$17, VMP$^+$18]. However, in its original form, DTW has a quadratic time complexity $\mathcal{O}(N^2)$, where $N \in \mathbb{N}$ is the length of the tested time series. Various methods are proposed to reduce DTW's

computation time [KeR05, SaC07, SiB16]. The proposed concept in [KeR05] reduces the time complexity of DTW to $\mathcal{O}(N)$ by the notation of upper and lower
envelopes, which provides the maximum number of warping. DTW's main disadvantage is that each data point of the tested time series must correspond to some
data point of the other time series, so even outliers should be considered in this
method [BeR14, BaR16]. In [SiB16], a prefix (suffix)-DTW (Ψ-DTW) is introduced
to handle DTW's problem regarding endpoints that exist before and after the segmented subsequence. These endpoints usually bear no information and are called
prefix and suffix. Ψ-DTW avoids undesirable matches at the endpoints by removing the alignment's criteria. Thus, on the contrary to DTW, there is no necessity
that each path must start and end in the first and last pairs of observations [SiB16].
The US distance [Keo03] allows a global scaling of time series [Fu11]. The major
problem of this method is the definition of the scaling factor. This usually represents
a time-consuming task and requires several trials. Fu et al. [FKL$^+$08] proposed a
new similarity measure, the Scaled and Warped Matching (SWM) [FKL$^+$08], which
benefits from both US and DTW methods. However, determining the scaling factor
and the time-warping constraint is a drawback of this method.
The SpADe measure [CNO$^+$07] applies a sliding window with a fixed size, allowing
both shifting and scaling on both axes in order to find patterns in time series
[WMD$^+$13]. This measure can be employed in streaming time series. Nevertheless,
several parameters must be adjusted and tuned for this method [BeR14].
Edit-based measures [Lev66, EsA12] or Levebshtein distances [Lev66] analyse two
time series based on the minimum number of required operations, e.g., insertions,
deletions, and substitutions, for transforming one series into another. Two time
series (or strings) of different lengths are aligned to become identical with the
smallest number of operations. Examples of these measures are Edit Distance
[Lev66] and the Longest Common SubSequence (LCSS) [PaD94]. The edit distance
method in its origin is employed for string data sets. An improved version of this
method is proposed in [Che05], which is called Edit Distance on Real Sequence
(EDR), and quantifies the distance between two numerical series. In contrast to
DTW and ED, this method can handle noisy data and outliers. EDR assigns
penalties when investigated parts of two time series do not match. It handles time
shifting in time series; however, it does not satisfy the triangle inequality and it is
not metric [WMD$^+$13, BeR14].
The idea behind LCSS [PaD94] is that time series have a fingerprint or distinct
segment that specifies them the best. The main problem of LCSS is its sensitivity to
noise. Moreover, a threshold and a warping parameter must be defined in advance.
Like DTW, this method's computational complexity is quadratic ($\mathcal{O}(N^2)$, where
$N \in \mathbb{N}$ is the length of time series) [WMD$^+$13].

3.4 Summary

This section has provided information about the general steps in time series motif
discovery. Segmentation, noise reduction, and normalisation are mostly performed

in the pre-processing step. In the representation phase, signals are transformed to approximate the data without losing the main characteristics. Among the stated methods, PAA [KCP$^+$01] and SAX [LKL$^+$03] are the most commonly-applied approaches in the domain of time series motif discovery. Comprehensive information and reviews of various representation methods can be found in [PaS17, HH19].

The last step is the similarity measurement. The choice of a proper similarity and distance measure strongly depends on the application and nature of the data [EsA12]. The family of L_p distances and elastic methods represent the most-utilised measures in time series motif discovery [LZP$^+$18, GaL18, FeN19]. A comparison of various distance measurement methods in diverse domains is available in [EsA12, WMD$^+$13, YoC15, RGC16, LFW$^+$19].

4 State of the Art in Time Series Motif Discovery

Time series are one of the most substantial parts of the world's supply of data, according to [Fu11], and they are part of several applications from diverse areas. This topic attracts several researchers from the domain of data and knowledge discovery. Data mining and pattern recognition tasks aim to provide information derived from time series [BeR14, BLB$^+$17, FaV17, AAJ$^+$19, AlA20]. Clustering, classification, query by content, and motif discovery belong to the class of pattern recognition and data mining tasks [BLB$^+$17, Tya17]. Although these tasks have a similar goal, namely detecting patterns, they differ in their nature and procedure for pattern discovery.

As stated in Chapter 1, motifs have been addressed in various literature as recurring patterns, frequent trends, approximately-repeated sequences, shapes, episodes, or frequent subsequences [CKL03, MIE$^+$07a, FGN$^+$09], all sharing the same goal. Regardless of these terms, the motif's definition can be categorized into the top K-frequent (Def. 2.8) [PKL$^+$02] and nearest-neighbour (Def. 2.10) [YKM$^+$07] motifs. As defined in Chapter 2, the top K-frequent motifs' definition is based on the concept of subsequence matching [PKL$^+$02]. The alternative definition, nearest-neighbour motifs, follows the concept of distance between two subsequences [YKM$^+$07].

Motif discovery algorithms differ in the way they tackle problems. These approaches can be analysed with regard to different aspects. As an example, motif discovery methods can be tailored to find exact or approximate motifs or detect motifs with fixed or various lengths. They are able to handle multivariate or univariate time series and execute in on- or off-line mode. Moreover, they can be examined based on representation or mapping methods [Mue14], similarity measures [Mue14], their robustness to noise, their ability to be invariant under affine transformations [ToL17b], or the number of required parameters.

The following sections provide an introduction to and a review of the state-of-the-art algorithms in time series motif discovery. It should be noted that the peer-reviewed work of the author published in [ToL17b] is integrated literally into this section.

4.1 Motif Discovery Algorithms

The algorithms in this section are classified based on their way of solving the problem of detecting motifs. For the sake of simplicity, when possible, the algorithms

© The Author(s), under exclusive license to
Springer-Verlag GmbH, DE, part of Springer Nature 2022
S. Deppe, *Discovery of Ill-Known Motifs in Time Series Data*, Technologien
für die intelligente Automation 15, https://doi.org/10.1007/978-3-662-64215-3_4

are presented in chronological order. Additionally, these algorithms are analysed based on their time complexity and ability to detect ill-known motifs (Def. 1.3).

Birth of Time Series Motif Discovery

Time series motif discovery has a short history so far, starting in 2002 with the introduction of the term motif by Patel et al. [PKL$^+$02]. The concept of this approach is to discover similar unknown subsequences based on the definition of top $K-$frequent motifs (Def. 2.8). Two subsequences match if the Euclidean distance (ED) [DeD09] between them is smaller than the pre-defined threshold $\tau \in \mathbb{R}^+$. This algorithm [PKL$^+$02] has weaknesses, such as inadequate scalability of finding motifs and erroneous behaviour with noisy data. The considerable time complexity ($\mathcal{O}(N^4)$) and pre-definition of parameters are other weak points of this method. Moreover, the similarity threshold (τ) and the statistical length of motifs (m) have to be provided for this concept. In order to speed up the performance of this approach, methods such as Enumeration of Motifs through Matrix Approximation (EMMA) [PKL$^+$02] and Efficient Motif Enumeration (MOEN Enum.) [MuC15] are proposed. These two approaches have profited from the symmetry and triangle inequality properties of the ED to speed up the search for motifs of equal length [ToL15]. The MOEN Enum. [MuC15] algorithm detects motifs of variable lengths, although this method does not compare such motifs with each other.

Chiu and Keogh [CKL03] applied the random projection algorithm [BuT01] to obtain motifs in DNA and protein sequences. This method detects the top K-frequent motifs. The novelty of this method is the discretisation step. The random projection method is inspired by concepts in biology and DNA sequences, whereby time series are transformed into discrete symbols. Discretisation is performed by the Symbolic Aggregation approXimation (SAX) [LKL$^+$03] algorithm. The authors of the random projection method stated that this method is robust in case of noise. However, several parameters must be defined, such as SAX parameters, the length of motifs, and the similarity threshold τ [LLC$^+$15]. Furthermore, as stated by Butler and Kazakov [BuK15], SAX has different deficiencies, like the assumption that every normalised time series is distributed normally [PKL$^+$02, LKL$^+$03,LKW$^+$07] or the symbols' equiprobability. SAX is applied in [WaM18] to detect motifs of equal length by means of the nearest neighbour classifier [Alp10].

Concept of Motif Balls

Liu et al. [LYL$^+$05] presented a motif ball concept similar to the top K-frequent motifs. In this approach, a time series $x[n]$ of length $N \in \mathbb{N}$ is treated as a data point in an N-dimensional space. Two time series are considered to be similar if the two corresponding N-dimensional data points are similar. The similar time series are grouped as N-dimensional data points in a ball with a radius $r \in \mathbb{N}$. Consequently, a motif is a dense ball with the highest number of data points in an N-dimensional space. Pre-definition of the length of motifs and the radius of motif balls r are the drawbacks of this method.

Detection of Inverse Motifs

The approach of Ferreira et al. [FAS+06] is also based on the top K-frequent motifs' definition. Nevertheless, subsequences that are altered by reflection mapping are also considered as motifs in this work. This method provides subsequences of various lengths, from the shortest subsequence of two samples up to the user-defined motif length m. After segmentation, Ferreira et al. [FAS+06] transform the subsequences into symbolic subsequences using SAX. Next, all transformed subsequences are compared with each other using the Pearson correlation coefficient [DeD09]. After measuring similarities, a bottom-up approach is begun by making clusters for each set of motifs, whereby motifs of the same length belong to the same cluster. The advantage of this approach is the ability to consider multivariate time series. Nevertheless, definitions for several parameters (besides SAX parameters) are required for this algorithm.

Concept of Grammar-based Motifs

The grammar-based algorithm applied by Lin and Li [LiL10] identifies top K-frequent motifs in medical data, such as electroencephalogram (EEG) signals. This method assigns a grammatical rule to each subsequence, converted to a symbolic subsequence using the SAX method. Each repeated SAX word is replaced by a grammatical rule that generates that subsequence [LiL10]. The top K-frequent motifs are the extracted rules that have the maximum number of occurrences. The grammar-based method helps to reduce the length of the sequence and summarize the structure of the data. This method is applied as a part of several rule discovery applications [SLW+14, SYCC+15]. Balasubramanian et al. [BWP16] employed this algorithm for detecting motifs in multidimensional time series gathered from a patient activity data set. This algorithm's two main advantages are the ability to handle data streaming and self-determination of the motif's length. Nevertheless, SAX parameters must be defined, and the performance of this method is affected by SAX problems [BuK15]. A similar approach is introduced in [GLR17], aiming to overcome the drawbacks of [LiL10, GLR17]. The method of Phein et al. [PNA19] discover motifs of variable-length by means of a suffix array and PAA-SAX representation method.

Detection of Exact Motifs

Most motif detection algorithms focus on discovering approximate motifs. Such motifs are not precisely similar. As a result, Mueen, Keogh, and Bigdely-Shamlo [MKBS09] proposed an algorithm for finding exact motifs, called Disk Aware Motif Enumeration (DAME).
DAME is based on four steps: a geometric view, a disk view, a projected view, and a search step. In the geometric view, each point of the data set is considered in a 2D space with a unique ID. In the disk view, the data points are stored in ascending

order in a disk block according to their distance to the randomly-selected reference point $r \in \mathbb{R}$. The projection view is considered as rotations of all of the points about the randomly-selected reference point r until they fall on a line, called the order line. The search procedure is a straight bottom-up approach. The search step for motifs begins with the smallest group size one and then continues by doubling the group size (i.e., 2, 4, 8, ...). Each group is compared with its neighbour group on the order line. At each iteration, disjoint pairs of the next groups are compared, and finally, all of the motifs are detected.

The principle of the order line speeds up the search procedure. However, this algorithm still has quadratic computational complexity. Another problem is that the reference point r either has to be determined in advance (by the user) or has to be chosen randomly. The location of this point affects the performance of this algorithm. As this method works directly on raw data, it is sensible to noise. Finally, in real-world applications, the chance of finding exact patterns or motifs is rare.

A similar method to DAME is proposed in [AlA20] for equal-length motif discovery. This method employs the concept of the order line, but the data is transformed by SAX in advance. Other methods, such as the Moen-Keogh algorithm (MK) [MKZ⁺09], also apply the same procedure as in [MKBS09]. The MK algorithm saves the best-so-far distance and updates the result, so there is no need to define a similarity threshold. In the MK algorithm, the number of reference points can be more than one, whereby the points are chosen randomly, which increases the speed of the algorithm. Like DAME, methods in [MKZ⁺09, AlA20] are sensitive to noisy data since it operates directly on the raw time series.

Self-determination of Motif's Length

Another major problem in motif discovery algorithms is the requirement of a pre-defined length for a motif. A general predicament is to provide the length of the motif, which usually needs expert knowledge. The proposed method in [LiL10] is able to tackle this problem. In [LLO12], an approach for visualization of variable-lengths motifs is proposed, called VizTree. The idea behind this method is based on grammar induction. VizTree employs an improved version of the algorithm in [LiL10]. Time series are transformed into SAX subsequences, and grammar rules are derived. Next, the most repeated rules are considered as motifs. No definition for the length of motifs is needed for this method, although the SAX parameters must be defined by the user. Noise sensibility and the inability to detect ill-known motifs (Def. 1.3) are other drawbacks of this method.

The proposed algorithm in [YSR⁺13] finds a proper length for motifs by applying the Minimum Description Length (MDL) method [Gru05]. MDL finds a subsequence of length $m \in \mathbb{N}$ as a hypothesis that best compresses the input time series. After detecting all of the subsequences, the similarity between them is measured by the ED. This method offers early termination, which is performed when a specific criterion is met. The user does not need to define any parameters in this work. However, in order to choose a subsequence of length m (hypothesis), it is necessary to have sufficient knowledge about the data.

Sivaraks and Ratanamahatana [SiR15] applied the same procedure to define a length for a motif via the MDL method. The similarity between motifs is measured by the Dynamic Time Warping (DTW) method [BeC94, Mül07]. This method is sensitive to noise. MDL is also employed in the work of Shokoohi-Yekta et al. [SYCC+15] to obtain the length of motifs, and later motifs are discovered by the MK algorithm [MKZ+09].

Motifs of Various Lengths

Different approaches aim to handle the problem of detecting motifs of variable lengths. Tang and Liao's approach [TaL08] begins by detecting motifs of the smallest length. Next, these motifs are concatenated to discover motifs of larger length. This approach is based on the random projection approach [CKL03]. The authors tackled the drawback of the random projection method, namely the inability to detect all occurrences of motifs. However, for this method, the SAX parameters must be provided.

Methods such as Variable Length Motif Discovery (VLMD) [NNR11] are proposed to solve the problem of motifs of different lengths. In each iteration of VLMD [NNR11], different lengths assign to a sliding window to segment the data into various subsequences. Next, motifs are detected in each iteration. Finally, motifs of the shortest to the most extended length are classified. This method finds motifs of various lengths, but it has enormous time complexity since it iterates the motif discovery algorithm for each assigned motif's length. Moreover, this approach results in a considerable amount of redundant motifs that are detected multiple times in each iteration.

Nunthanid and his group [NNR12] presented a non-parametric method, called k-Best Motif Discovery (kBMD), which tries to handle the problem of detecting motifs of variable lengths. In the first step, all subsequences with variable lengths are analysed before similar motifs are grouped. Subsequently, a motif in each group is selected by the minimum normalised ED from each motif group, whereby motifs that are too long are discarded. Finally, motifs at a similar location are grouped as best motifs. A sorting algorithm finds the longest motifs of each group and ranks them accordingly. Berlin and Van Laerhoven [Bev12] applied a Sliding Window And Bottom-up (SWAB) algorithm [KCH+04] to generate subsequences of various lengths. Next, all subsequences are transformed into symbolic sequences and then set into a suffix tree. Motifs are detected by searching the suffix tree to a certain depth and accumulating those motifs that occur at least two times. Although this method detects motifs of variable lengths, its disadvantages are its noise sensitivity and the pre-definition of the symbolic representation method's parameters.

In [GSI+14], Gulati and his colleagues introduced a method to find short-time melodic motifs in audio signals. After segmenting the whole signal using a sliding window, the similarity between subsequences is measured by different versions of DTW. This method can successfully detect and compare motifs of variable lengths. However, many parameters, such as window length and the number of pitch samples, are defined empirically. Additionally, this method is sensitive to noise [Fu11]. Liu et al. [LLC⁺15] suggested a method to find motifs of variable lengths in healthcare applications. The method is based on the random projection method proposed in [CKL03].

A similar approach is proposed by Mohammad and Nishida [MoN16], based on the extended versions of the MK algorithm [MoN14]. Zan and Yamana [ZaY17] to detect motifs of variable length within two steps. In the first step, after segmenting the data into subsequences, they are transformed by SAX to string arrays of symbols. The position of each subsequence is preserved in an index array. Subsequently, not-overlapping candidate motifs are discovered from the transformation domain. In the second step, these candidate motifs are verified with real-valued time series using DTW to guarantee the quality of discovered motifs. One limitation of this method is determining several parameters required in the sliding window and SAX methods. Being sensitive to noise is another drawback of this method.

Gao and Lin [GaL18] introduced a HierarchIcal based Motif Enumeration method (HIME) for variable lengths motifs discovery. This method applies SAX and an induction graph in its representation step. After transforming the subsequences to SAX words, an induction graph is made where each node of the graph represents a subsequence. This induction graph enumerates the motif candidates during motif discovery. One main advantage of this method is its scalability (comprising millions of data points). However, SAX parameters must be provided in advance, and this method is unable to discover ill-known motifs.

In [BrB18], a method based on the Self-Organising Map (SOM) is given. This method utilises sliding windows of different lengths to obtain variable-length subsequences, after which these subsequences are forwarded to a self-organising map to detect motifs. Several parameters must be provided and initialised for this method. The values of these parameters are gathered from a sensitivity analysis applied in [AnA16] to determine the impact of the parameter values on the model response and select the most appropriate parameter values. A similar method to this approach is presented in [InH20] which applies the DTW instead of the ED.

Madrid et al. [MIM⁺19] proposed a method based on the concept of the matrix profile for detecting and visualising motifs of various lengths. This method is sensitive to noise since it employs the ED and cannot detect ill-known motifs. Moreover, the similarity threshold for detecting motifs must be set by a domain expert. Several versions of this method are introduced in [ImK19, KGK19] for detecting semantic motifs in signals analysing human behaviour or searching the matrix profile [ZYZ⁺20]. Variable-length motif discovery applying deep neural networks is given in [RCL⁺20], which also utilises the ED. For this method, network parameters must be defined in advance.

Motifs with Local Variability in Time Domain

Detecting motifs with different local variability in the time domain is another research problem in motif discovery. To tackle this problem, Yankov et al. [YKM⁺07] have introduced a new definition for the term motif, namely the nearest neighbour motifs (cf. Def. 2.10). In order to determine motifs with local variability in the time domain, the Uniform Scaling (US) measure is applied in this method. A disadvantage of this method is that the motif's length must be provided in advance. Moreover, many examinations must be performed to find the proper scaling factor for the uniform scaling method. Nevertheless, by introducing the new definition for motifs, there is no need to assign the similarity threshold parameter, as it is the case in the top K-frequent motifs discovery approaches.

The MK algorithm [MKZ+09], which detects exact motifs, also performs by detecting the nearest neighbour motifs. The DTW measure applied in numerous time series motif discovery applications is partly able to handle the problem of patterns having stretches in the time axis.

Silva and Batista [SiB18] have recently proposed a method based on the approach in [ZYZ+18]; nevertheless, instead of ED, the prefix- and suffix- invariant DTW (ψ-DTW) [SiB16] distance is applied. The distance measure ψ-DTW quantifies the matching of two subsequences under DTW, ignoring up to $r \in \mathbb{N}$ endpoints. This method has disadvantages, such as the discovery of fixed length motifs and sensitivity to noisy data.

Motif Discovery in Multi-dimensional Data

An approach to detect motifs in multi-dimensional time series data is presented in [TIU05]. In this method, the Principal Component Analysis (PCA) [Pea01, Alp10] converts the multi-dimensional data to a one-dimensional signal. The one-dimensional signal is sent to the random projection algorithm [CKL03]. Finally, motifs are extracted by the MDL algorithm [Gru05]. This method finds motifs of variable lengths in different dimensions. Applying SAX as one part of this method makes it sensitive toward noise and parameters' definition. Furthermore, the results of this method mainly depend on the type of application and data [MSE+06]. A similar method to find motifs in multi-dimensional data is proposed by Minnen et al. [MIE+07a]. Here, SAX [LKL+03] is employed in the pre-processing step to extract and convert subsequences into SAX symbolic sequences. Subsequently, the random projection method [CKL03] is performed to discover equal-length subsequences in linear time. One problem regarding the random projection algorithm is that it cannot determine the motifs' dimensionality. Therefore, the random projection is applied to each dimension separately, which increases the time complexity of the method. Another limitation of this method is the definition of several parameters: SAX parameters, the length of motifs, and the similarity threshold τ.

Vahdatpour, Amini, and Sarrafzadeh [VAS09] employed the same procedure as Minnen et al. [MIE+07a]. However, they used the method proposed by Yankov et al. [YKM+07] to detect motifs with local variability in each dimension. After detecting motifs in a single dimension, a graph clustering approach constructs multi-dimensional motifs by combining the previously detected motifs. In contrast to the method of Minnen et al. [MIE+07a], this method determines motifs that are not synchronous across all time dimensions. Identifying SAX parameters is one of the problems of this method.

The proposed method in [BWP16] distinguishes motifs in multi-dimensional signals by adapting the method in [LLO12]. After investigating all motifs, motifs in different dimensions are grouped based on their timestamps. As this approach is based on the SAX representation method, it is sensitive to noise. Besides, values for several parameters must be assigned in advance.

Yeh et al. [ZYZ+18] proposed a multi-dimensional motif discovery approach based on a matrix profile and a distance measure called MASS [MZY+17]. Several improvements of matrix profile based methods are proposed [LZP+18, ImK19, ZYZ+20, RVR+20]. The authors claim that the algorithm is parameter-free; however, the length of the motifs must be provided in advance. Additionally, this method only identifies motifs of equal length.

Multi-Scale Motifs' Detection

Li et al. [LVK+04] employed a clustering approach at lower scales of wavelet decomposition for identifying top K-frequent motifs. The authors stated that any wavelet decomposition is applicable in this approach, although the Haar wavelet is performed due to its simplicity. After decomposing the time series $x[n]$ of length $N \in \mathbb{N}$ into $s \in \mathbb{N}$ scales satisfying $1 \leq s \leq \log_2(N)$, the K-Means clustering algorithm [Alp10] is applied to the coefficients in each scale to find motifs. This method is an any-time procedure. Such methods have a best-so-far answer which allows the user to examine this answer at any time. The user is able to either terminate the algorithm or run it for the complete procedure. This property reduces the time complexity of the K-means clustering and provides more flexibility. Regardless of its advantages, the number and the centre of the clusters (at the first run) as well as the length of motifs must be provided.

Castro and Azevedo proposed the Multi resolution Motif (Mr.Motif) algorithm [CaA10] which employs iSAX [ShK08] to provide an amplitude multi-resolution motif detection method. iSAX is an extended version of SAX and represents a SAX word in different resolutions. The term resolution in Castro and Azevedo's method means the size of the symbols for a SAX word. Thus, iSAX allows different symbol sizes (resolution) for the same word by dividing the amplitude of the time series into smaller sections, where each section is considered as a resolution. The iSAX method is typically used as an indexing tool for large time series; however, in [CaA10], this method is applied as a representation tool. Mr.Motif, which detects motifs of equal length, belongs to the state-of-the-art algorithms; nevertheless, its performance is affected by the SAX mapping method. Moreover, this method provides poor results in the case of noisy data.

In order to identify motifs in multiple temporal scales, Vespier et al. [VNK13] employed MDL [Gru05]. This method determines the motifs of variable lengths in each scale. However, several repetitive motifs are obtained since this method cannot compare the detected motifs with each other. Motifs in different frequency scales are detected in personal lifelogging time series by applying maximum overlap discrete wavelet transform [LCG+16]. In this method, data are transformed into different wavelet scales before SAX converts each scale's coefficients into symbolic sequences. Finally, motifs are detected in each scale by the minimum description length method. The definition of SAX parameters and the problems accompanying it [BuK15] are the main disadvantages of this method.

4.1.1 Time Complexity

One of the most critical characteristics of motif discovery algorithms is time complexity. Consequently, this issue is considered in this state of the art.

The first time series motif discovery method [PKL$^+$02] has the time complexity of $\mathcal{O}(N^4)$, where $N \in \mathbb{N}$ is the length of a time series. This method's time complexity is halved by proposed methods such as EMMA [PKL$^+$02] and MOEN Enum. [MuC15]. The proposed method of Chiu et al. [CKL03] has a quadratic time complexity which depends on the length of the SAX word.

The time complexity of the clustering-based approach [LVK$^+$04] is $\mathcal{O}(kNrD)$, where k, D, and r are the number of clusters, time series, and iterations, while N is the length of the time series ($k, D, r, N \in \mathbb{N}$). A similar any-time method is proposed by Serrà and Arcos [SeA16], which finds non-overlapping motifs based on the particle swarm optimisation and performs in $\mathcal{O}(N^2 w^2)$, where $N, w \in \mathbb{N}$ are the length of the time series and sliding window, respectively.

The motif ball concept [LYL$^+$05] has the time complexity of $\mathcal{O}(N^2)$, where $N \in \mathbb{N}$ is the length of the time series. The method introduced in [TIU05] determines motifs in multidimensional data and has quadratic time complexity. Anh [Anv16] has recently accelerated the execution of this method and provided an efficient implementation.

The method of Yankov et al. [YKM$^+$07] has time complexity equal to $\mathcal{O}(cN^2 m)$ for some constant $c > 1$, time series length N, and motif's length m. The time complexity of the approach of Minnen et al.'s [MIE$^+$07a] is quadratic since the DTW is applied in this method. Vahdatpour [VAS09] applied the same procedure as [MIE$^+$07a] with the same quadratic time complexity. The variable-length motif discovery approach given in [TaL08] performs in $\mathcal{O}((|sM|^2 - |sM|)/2)$, where sM is a set of all identified motifs.

The MK algorithm [MKZ$^+$09] for the exact motif discovery is faster than the standard brute-force algorithm and has a time complexity of $\mathcal{O}(nR)$. The number of time series is given by n, and R is the number of user-selected reference points ($n, R \in \mathbb{N}$). DAME [MKBS09] and its improved version [AlA20] perform in $\mathcal{O}(N^2)$, where N is the length of the time series. Lin's grammar-based method [LiL10] detects motifs in linear time. Applying this grammar-based method in [LLO12] leads to the computation complexity of $\mathcal{O}(w)$, where w is the length of SAX word. Similar approaches are employed in [Bev12, GLR17, PNA19] with the time cost equal to $\mathcal{O}(w \log w)$.

Castro and Azevedo's [CaA10] amplitude multi-resolution method performs in linear time.

The kBMD [NNR12] and VLMD [NNR11] methods both apply the MK algorithm [MKBS09] to detect motifs. According to the authors [NNR11], these methods have a time complexity of $\mathcal{O}(NmnR)$, where N is the length of the time series, m is the motif length, and the parameters n and R are the same as in the MK algorithm. It must be noted that these methods must be performed $k \in \mathbb{N}$ times to detect k types of motifs of various lengths.

The presented method by Yingchareonthawornchai et al. [YSR⁺13] has $\mathcal{O}(m^2N^2)$ time complexity, where N and m are the lengths of time series and motifs ($N, m \in \mathbb{N}$).

The time complexity of the method in [VNK13] is $\mathcal{O}(N \log_2^N + |\mu|(\log|\mu| + |c|N))$, where N is the length of time series, μ is the number of symbolic representation strings, and c is the number of candidate motifs ($N, \mu, c \in \mathbb{N}$) [VNK13].

Sivaraks' and Ratanamahatana's [SiR15] method is based on MDL and DTW and it has a quadratic time cost. The variable lengths motif discovery proposed in [LLC⁺15] has quadratic time complexity. Approaches in [BWP16, InH20] detect motifs in each dimension using the method proposed in [LLO12] with the time complexity $\mathcal{O}(k2^k)$, where $k \in \mathbb{N}$ is the number of candidate motifs.

The proposed approach in [LiL19] and the multi-scale method in [LCG⁺16] has a linear time complexity $\mathcal{O}(N)$. All three steps in this method have a linear time complexity: transforming data into different scales is undertaken in $\mathcal{O}(N)$, symbolic representation is obtained in $\mathcal{O}(w)$, and detecting the suitable length for the motifs performs at $\mathcal{O}(|c|N)$. The length of the time series, the symbolic word size, and the number of the candidate motifs are given by N, w, and $|c| \in \mathbb{N}$, respectively.

Zan' and Yamana's method [ZaY17], which comprises two stages, determines the variable-length motifs. In the first stage, this method has a time complexity of $\mathcal{O}(|SA|^2)$, where $|SA|$ is the length of the string array. In the second step, this method performs in $\mathcal{O}(N^2)$, where N is the length of the time series. Other approaches for variable-length motif discovery, like [RCL⁺20], have a quadratic time complexity.

The authors in [SiB18] proposed a method similar to [YKM⁺07] that improves the time complexity to $\mathcal{O}(N^2)$ with $N \in \mathbb{N}$ equals to time series' length. The matrix profile method of [ZYZ⁺18, MIM⁺19, ImK19, ZYZ⁺20] performs in $\mathcal{O}(N \log N)$.

Table 4.1 summarises the computational complexity of the aforementioned algorithms.

Table 4.1: A review of the time complexity of the state-of-the-art algorithms.

Method	Time Complexity	Parameters
[PKL⁺02]	$\mathcal{O}(N^4)$	N time series' length
[CKL03]	$\mathcal{O}(w^2)$	w length of SAX word
[LYL⁺05]	$\mathcal{O}(n^2)$	n number of time series
[FAS⁺06]	$\mathcal{O}(m^2)$	m motif's length
[YKM⁺07]	$\mathcal{O}(cN^2m)$	c constant; N time series' length; m motif's length
[MIE⁺07a]	$\mathcal{O}(m)$	m motifs' length
[TaL08]	$\mathcal{O}((|sM|^2 - |sM|)/2)$	sM a set of all motifs

[MKZ$^+$09]	$\mathcal{O}(nR)$	n number of time series; R number of reference points
[MKBS09, GSI$^+$14] [MuC15, SiR15] [SiB18, SiB18]	$\mathcal{O}(N^2)$	N time series' length
[LiL10, LLO12]	$\mathcal{O}(w)$	w length of SAX word
[NNR11, NNR12]	$\mathcal{O}(NmnR)$	N time series' length; m motif's length; n number of time series; R reference points
[Bev12, PNA19]	$\mathcal{O}(w \log w)$	w length of the symbols
[YSR$^+$13]	$\mathcal{O}(m^2 N^2)$	N and m length of time series and motifs
[VNK13]	$\mathcal{O}(N \log_2^N + |\mu|(\log |\mu| + |c|N))$	N length of time series; μ number of symbolic representation strings; c number of candidate motifs
[LLC$^+$15, LCG$^+$16] [LiL19]	$\mathcal{O}(N)$	N time series' length
[SYCC$^+$15]	$\mathcal{O}(N \log N + |maxlag|)$	N time series' length; $maxlag$ steps between two subsequences
[BWP16, InH20]	$\mathcal{O}(k2^k)$	k number of candidate motifs
[MoN16]	$\mathcal{O}(Rn \log n)$	n number of time series; R number of reference points
[GLR17]	$\mathcal{O}(2 \log(w_{\max}) + 7s)$	s length of the sliding window; $w_{\max}$ maximum length of SAX words
[ZaY17, RCL$^+$20]	$\mathcal{O}(|SA|^2)$	$|SA|$ length of string array
[GaL18]	$\mathcal{O}(2 \log(w_{\max}) + 7s)$	s length of the sliding window; $w_{\max}$ maximum length of SAX words
[ZYZ$^+$18, ImK19] [MIM$^+$19, ZYZ$^+$20]	$\mathcal{O}(N \log N)$	N time series' length

4.1.2 Detecting Ill-Known Motifs

Real-world data are affected by various types of distortions. Noise, phase differences, affine transformations such as time or amplitude translations, scaling, and reflections are examples of such distortions. Consequently, in order to detect motifs under these circumstances, three types of approaches are available.

One possibility is to remove them in the pre-processing step. Batista [BKT$^+$14] provides a review of methods applied in the pre-processing step that deal with these problems. The problem of amplitude differences of two similar signals is tackled by normalising both data [KeK03]. As an example, two signals with similar shape

measured on different scales, e.g. Celsius and centigrade, cannot be well matched
since they have different amplitude scales. Noise is one of the most common distortions in data, which is usually removed in pre-processing steps. Nonetheless,
motif discovery methods such as [Bev12, YSR$^+$13, MuC15, GaL18, SiB18, WaM18]
are still sensitive to noise.

The second type of these methods conducting distortions in motif discovery focuses
on the similarity measurement step. The choice of distance measure is very vital
when working directly with the raw data. If the distortions are not removed in
advance, then the distance measure must be robust and invariant [EsA12, Spi15]
to, e.g., time and amplitude translations, rotations, scaling, or reflections of motifs.
Employing similarity measures that are not resistant to such mappings affect the
performance of the motif discovery methods. As an example, ED is one of the most
commonly-applied distance similarity measures [Mue14]. However, this measure is
shift variant and sensitive to outliers. Local misalignments are another issue that
occurs in time series. For instance, a small fluctuation of the speakers' tempo
recorded in speech signals may challenge identifying similar contents [FKL$^+$08].
Methods that apply the DTW measure are able to handle local scaling in the data
[GSI$^+$14, SiR15]. The US measure applied in [YKM$^+$07, VAS09] enables detecting
motifs with local scaling.

The targets of the third class of approaches are the representation or mapping methods that are invariant under the aforementioned transformations (affine mappings).
As an example, time series representation methods such as SAX, Fourier transform,
and (some) wavelet transforms are shift invariant. Consequently, methods such as
EMMA [PKL$^+$02], MOEN Enum. [MuC15], random projection [CKL03], Mr.Motif
[CaA10], and approaches in [LiL10, LLC$^+$15, LCG$^+$16, GLR17, ZaY17, GaL18] are
invariant to time translations.

As explained before, changing the length of motifs results in detecting more than
one type of motifs. Thus, besides being invariant to different affine mappings, a motif discovery algorithm must be able to detect motifs of variable lengths [ToL17b].
Most of the stated motif discovery algorithms are able to detect motifs with a pre-
defined fixed length, although this length has to be provided for several algorithms
[YSR$^+$13, GSI$^+$14, LCG$^+$16, SiB18, ZYZ$^+$18] in advance. The MDL method [Gru05]
provides a proper length for the motif but requires sufficient knowledge about the
data and the application. The detection of motifs with variable lengths is solved
by methods such as [LiL10, NNR12, NNR11, PNA19, InH20]. Other methods, such
as [HSYP$^+$13, GSI$^+$14, ToL14, ToL15, LLC$^+$15, RCL$^+$20], also employ the same procedure to detect motifs of variable lengths and size, namely by performing sliding
windows of different sizes. These methods determine motifs with variable lengths.
However, they [MKZ$^+$09, LiL10, HSYP$^+$13, SiR15, MIM$^+$19] cannot compare these
motifs with each other or find motifs that are altered by stretch or squeeze mappings.

To sum up, in motif discovery approaches, the problems of noise and scaling are
mostly tackled in the pre-processing step. Time and amplitude translations and
local misalignments are considered in either the representation or the similarity
measurement step. None of the methods determines motifs that are altered by

stretch or squeeze mappings.

4.2 Research Gaps

Time series motif discovery is a growing and appealing field that has raised challenges and research issues. In the last decade, several approaches have been proposed and performed to tackle the problem of this domain. These algorithms are analysed in Table 4.2 based on their requirements or abilities, such as the definition of motifs' length in advance, the applied representation method and distance measure, the capability to detect motifs of variable lengths, the property of multi-resolutions analysis (deriving motifs at different resolutions), and the ability to determine motifs directly in the time or frequency domain. Additionally, the time complexity of these methods is also benchmarked.

Table 4.2: Related work in time series motif discovery. Methods are analysed based on various criteria, such as providing the motif's length in advance, applied representation methods and distance measures, motif analysis in the frequency domain, ability to determine equal- and variable-length motifs, discovery of motifs in different resolutions, and having a linear time complexity. If a method fulfils any of these criteria, then a + is assigned to the method [ToL17b].

	Motif's length	Representation method	Frequency domain	Similarity measures	Variable length	Multi-Resolution	Linear time complexity
[PKL+02]	+	SAX	-	ED	-	-	-
[CKL03]	+	SAX	-	ED	-	-	-
[LVK+04]	+	Wavelets	ED	+	-	-	-
[TIU05]	+	SAX	-	-	-	-	-
[FAS+06]	+	SAX	-	Corr.	+	-	-
[YKM+07]	-	SAX	-	US	+	-	-
[TaL08]	-	-	-	ED	-	-	-
[MKZ+09]	+	-	-	ED	-	-	+
[LiL10]	+	SAX	-	ED	+	-	+
[CaA10]	-	iSAX	-	ED	-	+	+
[NNR11]	-	-	-	ED	+	-	-
[LLO12]	-	SAX	-	-	-	-	-
[Bev12]	+	SAX	-	-	-	-	+
[YSR+13]	+	MDL	-	ED	+	-	-
[GSI+14]	-	-	-	DTW	+	-	-

[LLC+15]	-	SAX	-	ED	-	-	-
[SiR15]	+	MDL	-	DTW	+	-	-
[MuC15]	+	-	-	ED	+	-	+
[LCG+16]	+	Wavelets & SAX	+	DTW	-	+	-
[GLR17]	-	SAX	-	-	+	-	-
[TrA17]	+	CCD	-	ED	-	-	-
[ZaY17]	-	SAX	-	DTW	+	-	-
[GaL18]	-	SAX	-	-	+	-	+
[BrB18]	+	SOM	-	W-ED	+	-	-
[WaM18]	+	SAX	-	ED	-	-	-
[SiB18]	+	-	-	DTW	-	-	-
[ZYZ+18]	+	-	-	MASS	-	-	-
[PNA19]	+	SAX	-	ED	-	-	-
[MIM+19]	+	-	-	ED	+	-	-
[AlA20]	+	SAX	-	ED	-	-	+
[InH20]	-	SAX	-	DTW	+	-	+
[RCL+20]	+	-	-	ED	+	-	-

Similar to any other scientific research field, aiming to find the solution for a specific problem often leads to other questions emerging. These open research issues, which are visible in Table 4.2, are:

- **Investigation of various time series representations**: Prior approaches in time series motif discovery analyse the data mostly in the time domain. Recently, SAX attracts researchers' attention due to being straightforward and fast [EsA12, Mue14, ToL17b]. However, as stated in [BuK15], SAX has various disadvantages. Investigations of methods that do not discretise the data may improve the motif discovery results. Methods such as Fourier transform [Fou22, Fou78] or wavelet transforms [Dau90, Mey93, Mal08] analyse the data e.g. in the frequency domain, or provide a comprehensive time-frequency representation. Although these transformations have proven their performance in several signal and image processing tasks [ZhG16, CLZ17, GCK+18], they are less prevalent in time series motif discovery. Consequently, methods are required that analyse the data in other domains.

- **Handling ill-known motifs**: Real-world data sets contain several distortions such as noise, time translations, reflections, and scaling, to mention a few. Several approaches have been proposed to overcome the stated problems in motif discovery such as normalisation, noise reduction [Mey93], and uniform scaling [YKM+07]. Nonetheless, methods that identify motifs altered by squeeze or stretch mappings or generally determine ill-known motifs (cf. Def. 1.3) are still required [ToL15].

- **Exploration of motifs with variable lengths**: The number of methods that are capable of detecting motifs of variable lengths has recently increased.

These methods mostly iterate the same algorithm for various motif lengths and then classify the detected motifs in different classes based on the length of motifs. This is a time-consuming task and results in a redundant amount of information. Thus, approaches are required that can tackle this missing point.

According to the state of the art, and to the author's best knowledge, none of the stated approaches are able to detect ill-known motifs. This contribution aims to fill the aforementioned research gaps (as explained in Sec. 1.3, data streaming is not included in the scope of this thesis) by a proposed solution that analyses time series in a comprehensive time-frequency resolution, handles ill-known motifs, discovers both exact and approximate motifs, and detects motifs of various lengths.

4.3 Summary

This chapter has provided a comprehensive review of the state-of-the-art approaches in time series motif discovery. The advantages and disadvantages of each approach have been outlined, and the scientific gaps have been identified accordingly. These gaps are mainly closed by the proposed approach in this work, namely KITE (ill-Known motIf discovery in Time sEries data).

5 Distortion-Invariant Motif Discovery

This section represents the core of this dissertation. According to the motivation given in Chapter 1, the review of the motif discovery algorithms, and the research gaps provided in Chapter 4, KITE's approach is proposed in this section. KITE aims to overcome the limitations of the existing methods regarding detecting ill-known motifs by means of its steps. KITE offers solutions for the following dilemmas: alterations occurred due to translation, proportion (uniform scaling), stretch , squeeze, mappings, definitions of motif's length, and selection of representative motifs.

This section introduces KITE's structure and guides through each phase. The information provided in Chapter 2 and Chapter 3 serves as the basis for this approach.

The main contributions of this chapter are the following:

- To identify ill-known motifs that are, e.g. transformed by affine mappings, superimposed with noise, and have variations in their lengths, KITE is proposed.

- KITE is based on the fusion of general procedures in pattern recognition and motif discovery. Apart from the representation and similarity measurement steps, feature extraction is added as a new step in the KITE approach. In pattern recognition, by employing features, which describe the data and assist the learning algorithm, patterns that are changed by various mappings or have different size are identified. Thus, this combination allows KITE to overcome the limitations of motif discovery domain (e.g. detection of variable-lengths motifs) with strength and merits of the other domain. To the best of the author's knowledge, there is no other motif discovery approach with such a structure. KITE's architecture is given in Sec. 5.1.

- KITE proposes two procedures for the specification of motifs' length. The first method intends to determine motifs of equal length, and the alternative one aims to detect motifs of various lengths. This issue is described in Sec. 5.2.

- The core of KITE, which is the base of an invariant representation method, is called *Analytic Complex Quad Tree Wavelet Packet transform* (ACQTWP). The ACQTWP has properties such as shift invariance, flexible time-frequency resolution, and approximate analytic representation of a signal. The ACQTWP transform and its properties are explained in Sec. 5.3.

S. Deppe, *Discovery of Ill-Known Motifs in Time Series Data*, Technologien
für die intelligente Automation 15, https://doi.org/10.1007/978-3-662-64215-3_5

- KITE's feature extraction step aims to fulfil the challenges that cannot be clarified in the previous steps (e.g. reflection mapping [Woo96, BuB09]). Several features have the potential to be considered as features. Nevertheless, to generalise KITE's approach, six features (the first four statistical moments as well as the maximum and minimum value of the wavelet coefficients phase) are assigned in this work. Further explanations and reasons regarding selecting these features are given in Sec. 5.4.

- KITE proposes a procedure to exclude *misleading* motifs from and to determine *representative* motifs among all detected motifs. Both types of motifs are defined in Sec. 5.6.

Furthermore, Sec. 5.7 provides an analysis of time complexity in each distinct step of KITE. The work of the author published in [ToL14, ToL15, TDDL16, ToL18] is integrated literally into this section.

5.1 KITE Architecture

Motif discovery and pattern recognition share the same goal: identifying (un)known patterns. A typical motif discovery procedure and its steps are depicted in Fig. 3.1 and explained in Chapter 3. Pattern recognition is based on the automatic discovery of routines (regularities or patterns) in data utilising various algorithms [Alp10]. Following the same goal is one reason why KITE combines the general procedures of motif discovery and pattern recognition. KITE's structure and its main phases are illustrated in Fig. 5.1. In addition to the standard steps of the motif discovery approach (pre-processing, representation, and similarity measurement), KITE's structure includes the feature extraction step. Feature extraction has several advantages: reducing the number of data, presenting distinctive aspects of the data, and increasing the performance [PNP+12]. Besides, by employing feature extraction, KITE aims to tackle challenges caused by mappings, such as reflections or variable-length motifs.

Motif discovery by KITE is performed on the input signals in five main steps, as explained in the following:

- First, the lengths of motifs are determined in the pre-processing stage. The length of motifs is either provided in advance (e.g. by an expert) and is set to a fixed value for equal-length motif detection, or KITE segments the input data into equal and variable-length subsequences to discover motifs. Thus, pre-processing outcomes are subsequences of equal or variable lengths which must be investigated in the next step.

- In the representation step, all subsequences are decomposed by the ACQTWP transform to detect ill-known motifs. ACQTWP reduces noise, is shift-invariant, and analyses signals of various lengths. Hence, ACQTWP aims to overcome the alteration issued by translation, stretch, squeeze mappings, and added noise.

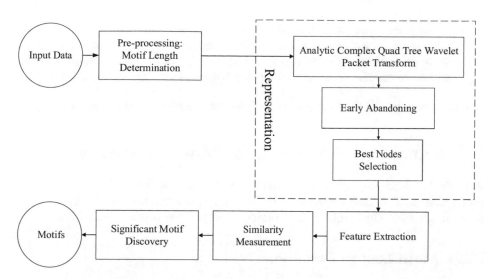

Figure 5.1: KITE structure: pre-processing, representation, feature extraction, similarity measurement, and significant motif discovery. The representation consists of three states: the Analytic Complex Quad Tree Wavelet Packet transform performs data transformation, early abandoning algorithm selects the most advantageous scale, and the most informative nodes are chosen for feature extraction.

Analysing and evaluating subsequences through the whole scales of ACQTWP leads to redundant information, and therefore, the "Best Basis" [Wic96, BaS08,GGS⁺09,WBK09,TDA15,CBA16,BPC⁺18,WRH18] are selected. The decomposition is conducted by the "*Early Abandoning*" algorithm. The criteria for stopping the decomposition and selecting a scale is based on each scale's shape preservation and information content. From the selected scale, the two most informative nodes are collected by the "*Best Node Selection*" (BNS) algorithm. These nodes are determined according to a criterion based on energy density and entropy [Sha01]. The wavelet coefficients of these nodes are forwarded to the feature extraction step.

- The number of data to be processed is reduced in the feature extraction step since instead of analysing the total amount of wavelet coefficients, features are extracted. Feature extraction assists KITE in discovering motifs which their dilemmas, such as length variation, are not handled in the representation step.

- The outcomes of the feature extraction are investigated in the similarity measurement step. Consequently, equal- and variable-length motifs are detected by measuring the similarity between their feature values by applying a distance measure (explained in Sec. 3.3). This stage results in the detection of all possible motifs.

- Finally, motif evaluation is performed to support the user. As mentioned in Chapter 4, motif examination is a time-consuming task mostly managed by an expert. For this reason, misleading motifs are excluded, while the representative motifs are selected from the the total amount of detected motifs.

After introducing KITE's design, each of these steps is clarified in the next sections.

5.2 Signal Pre-Processing for Motif Discovery

A typical obstacle regarding motif discovery methods is the definition of a proper motif length. This task depends on the tested application and needs expert knowledge in general. This problem is explained and tackled in the next section.

5.2.1 Motif Length Definition

Determining lengths for motifs is a challenging task not only for non-experts but also for experts. As an example, the domain experts may perform some trials to figure out the motif length. Two options for the length of motifs are considered in this work: pre-defined and non-defined motif length.

Definition 5.1 (Pre-defined motif length). Given a pre-defined motif length $l_d \in \mathbb{N}$, a sliding window (defined in Def. 2.5) $w[n]$ of length l_d extracts all possible subsequences from the time series $x[n]$.

In Def. 5.1, mostly an expert provides the (pre-defined) motif length in advance for equal-length motif detection. The procedure of deriving non-defined motif length follows the same manner.

Definition 5.2 (Non-defined motif length). Given a time series $x[n]$ of length $N \in \mathbb{N}$, a sliding window $w[n]$ of length l_{nd} for all $2 \leq l_{nd} \leq \frac{N}{2}$ divides the time series $x[n]$ into segments of length l_{nd}. The shortest segment contains two samples, and the length of the most extended segment is equal to half of the time series' length.

Due to lack of information on motifs' length, Def. 5.2 allows obtaining subsequences of equal and variable lengths from a time series without defining the length of motifs in advance. Motif length definition is a fundamental part since small differences in the length of motifs result in discovering diverse motifs. The following example explains this issue.

Example 2. Consider the signal given in Example 1. Setting the motifs' length to 800 leads to the detected motifs (m_1, m_2) in Fig. 5.2 (b).
By changing this length to 904, besides motifs (m_1, m_2), KITE also discovers motifs (m_3, m_4) as illustrated in Fig. 5.2 (c). This is not possible for most motif discovery algorithms, since they detect only motifs of the fixed length.

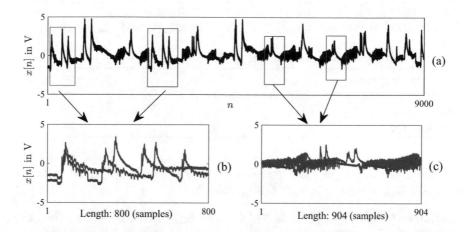

Figure 5.2: Changing the motif's length results in various motifs. Two different pairs of motifs with lengths of 800 and 904 samples are detected.

Thus, by varying the length of motifs, different types of motifs are distinguished. Depending on the application, it may be necessary to obtain subsequences which overlap each other. For both definitions above, it is possible to provide an overlapping degree $O_{\mathrm{d}} \in \mathbb{R}$. The overlapping degree between two subsequences is defined by:

Definition 5.3 (Overlapping degree). The overlapping degree $O_{\mathrm{d}} \in \mathbb{R}$ between two subsequences $s_{i,p} = (s_i, ..., s_{i+p-1})^{\mathrm{T}}$ and $s_{j,q} = (s_j, ..., s_{j+q-1})^{\mathrm{T}}$ of lengths p, $q \in \mathbb{N}$ and the beginning positions $i, j \in \mathbb{N}$ is computed by

$$O_{\mathrm{d}} = \frac{|s_{i,p} \cap s_{j,q}|}{\max(p, q)},$$

where $\cap$ is the intersection of two subsequences, and $O_{\mathrm{d}} \in [0, 1]$.

If $O_{\mathrm{d}} = 0$, subsequences do not overlap, and $O_{\mathrm{d}} = 1$ indicates the complete overlapping between two subsequences. The algorithms for pre-defined and non-defined motif length with an overlapping degree are given below:
Algorithm 1 divides the time series $x[n]$ of length $N \in \mathbb{N}$ into segments of equal length.

Remark 1. Algorithm 1 results in $N_{\mathrm{s}} = \lceil \frac{N-(l_{\mathrm{d}}-1)}{\lfloor l_{\mathrm{d}}-(l_{\mathrm{d}} \cdot O_{\mathrm{d}}) \rfloor} \rceil$ subsequences. If $l_{\mathrm{d}} - (l_{\mathrm{d}} \cdot O_{\mathrm{d}}) < 1$, then $\lfloor l_{\mathrm{d}} - (l_{\mathrm{d}} \cdot O_{\mathrm{d}}) \rfloor$ must be considered as 1 (preventing dividing by zero).

Algorithm 2 provides various subsequences from the smallest size of two samples to the largest size $N/2$. Besides dividing the time series $x[n]$ into all possible lengths, it is possible to set boundaries for the length of the motifs (the sliding window $w[n]$) in Algorithm 2 by assigning parameters $l_{\min}$ and $l_{\max} \in \mathbb{N}$ to detect motifs with

---

Algorithm 1 Sliding window with pre-defined length

Input: signal or time series $x[n]$, pre-defined length of motifs l_d, overlapping degree O_d

Output: subsequences $\mathbf{S}_i[n]$ of equal length

1: $N = \mathrm{length}(x)$;
2: **for** $i = 1 : (l_\mathrm{d} - (O_\mathrm{d} \cdot l_\mathrm{d})) : N - (l_\mathrm{d} - 1)$ **do**
3:     $\mathbf{S}_i = x[i : i + (l_\mathrm{d} - 1)]$;
4: **end for**

---

lengths given in the provided range. For a signal $x[n]$ of length $N \in \mathbb{N}$, variables l_min and l_max are set to 2 and $N/2$ by default.

Remark 2. Let $l_\mathrm{nd} = \{2, 3, ..., N/2\}$, where $l_{\mathrm{nd}_1} = 2$, $l_{\mathrm{nd}_2} = 3, ...$, then the total number of subsequences obtained by executing Algorithm 2 is given by

$$N_\mathrm{s} = \sum_{i=1}^{Card(l_\mathrm{nd})} \lceil \frac{N - (l_{\mathrm{nd}_i} - 1)}{\lfloor l_{\mathrm{nd}_i} - (l_{\mathrm{nd}_i} \cdot O_\mathrm{d}) \rfloor} \rceil.$$

If $l_{\mathrm{nd}_i} - (l_{\mathrm{nd}_i} \cdot O_\mathrm{d}) < 1$, then $\lfloor l_{\mathrm{nd}_i} - (l_{\mathrm{nd}_i} \cdot O_\mathrm{d}) \rfloor$ must be considered as 1. The size of the set l_nd is obtained by $Card(l_\mathrm{nd})$.

Algorithm 2 Sliding window with non-defined length

Input: signal or time series $x[n]$, overlapping degree O_d, (parameters l_min and l_max are by default 2 and $N/2$)

Output: subsequences of variable length $\mathbf{S}_{j,i}[n]$

1: $N = \mathrm{length}(x)$;
2: **for** $j = l_\mathrm{min} : l_\mathrm{max}$ **do**
3:     **for** $i = 1 : (j - (O_\mathrm{d} \cdot j)) : N - (j - 1)$ **do**
4:         $\mathbf{S}_{j,i} = x[i : i + (j - 1)]$;
5:     **end for**
6: **end for**

To sum up, in the first step of KITE, the input time series is divided into subsequences of equal or variable lengths by executing either Algorithm 1 or 2. These segmented subsequences are mapped by an invariant time series representation in the second step, explained in the next section.

5.3 Invariant Time Series Representation

As explained in Sec. 2.4, wavelet transformations are powerful representation or mapping tools that are also applied in this work since they can reduce noise [SÇŞ15, HaB15, RMB16, UmY18] and detect patterns that are transformed by e.g.

translation, stretch or squeeze mappings [BuM80, Wol88, GGS⁺09], and analyse the data in a flexible time-frequency resolution. Among several wavelets transforms, the dual-tree complex wavelet transform (DTCWT) [Kin01] was the first candidate to be applied in the KITE method due to its properties (cf. Sec. 2.4). However, it turns out that the DTCWT is not shift invariant and has other limitations like deficient decomposition (cf. Sec. 2.4) [ToL15]. The straightforward approach to overcome the DTCWT's deficient decomposition is to apply wavelet packets which form a Hilbert pair (cf. Def. 2.12). The outcome is a wavelet transformation comprising two wavelet packet trees working in parallel, *WPT A* and *WPT B*, where WPT A represents the real part and WPT B the imaginary part of a signal [BaS08]. The structure of this wavelet transform in two scales is illustrated in Fig. 5.3.

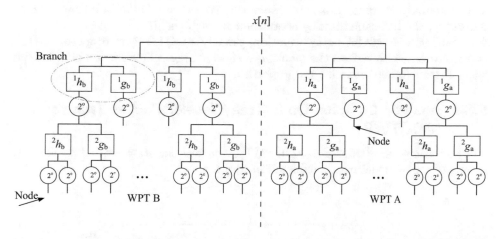

Figure 5.3: Filters $^s g_a$ and $^s h_a$ are the low- and high-pass filters of the WPT A, while the filters of WPT B are given by $^s g_b$ and $^s h_b$. $\downarrow 2^e$ and $\downarrow 2^o$ are the even and odd downsamplers.

This wavelet transformation achieves a shift-invariant transformation by following the proposed procedure: Instead of one downsampler after each low- and high-pass filtering, two downsamplers are performed to select a non-shifted and a shifted version of the input signal in each scale [ToL15]. The reason for this is that a signal translation of S samples is bounded by $mod(S, 2)$ in each scale induced by downsampling (factor 2 downsampling) [GGS⁺09, ToL15].

In Fig. 5.3, the downsamplers are depicted by $\downarrow 2^e$ and $\downarrow 2^o$ for the even and odd down-sampling, respectively. The two depicted terms in Fig. 5.3, *branches* and *nodes*, are defined as follows:

Definition 5.4 (Branches and Nodes). A branch splits the signal into two parts in the proposed wavelet tree after applying two filters. Each branch consists of a low- and a high-pass filter, as well as an even and odd downsampler. A node is constructed after each decomposition. Thus, each branch of the given schema in Fig. 5.3 has four nodes.

As an example, the second scale consists of 8 branches and 32 nodes.

Signal $x[n]$, given in Fig. 5.3, is decomposed by the low- and high-pass filters ${}^s g_{\mathrm{a,b}}$ and ${}^s h_{\mathrm{a,b}}$, where $s \in \mathbb{N}$ is the number of scales. Indices a,b refer to wavelet packets A and B.

The filters of this wavelet transformation are the same as in the DTCWT, given in Def. 2.13 and Def. 2.14, since these filters satisfy the conditions required for having an analytic and complex wavelet transform (cf. Sec. 2.4.1). Thus, applying these filters must result in an analytic transformation, where the response of each branch of the WPT A and the corresponding branch of the WPT B forms a Hilbert pair (cf. Def. 2.12). However, it turns out that applying the same filters in each branch and scale as in the straightforward structure, depicted in Fig. 5.3, results in a non-analytic transformation. Since this structure adds an additional phase difference, the half-sample delay condition cannot be held.

Consequently, a wavelet transformation, namely ACQTWP, is proposed in this contribution, which tackles the mentioned shortcomings. This section clarifies this wavelet transform along with its properties.

5.3.1 Analytic Complex Quad Tree Wavelet Packet Transform (ACQTWP)

The analytic wavelet transform ACQTWP overcomes the above-stated problem by means of a new structure, as depicted in Fig. 5.4.

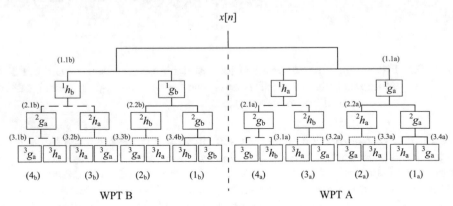

Figure 5.4: ACQTWP transform. Solid lines depict the regular pairs of filters $({}^s g_{\mathrm{a}}, {}^s h_{\mathrm{a}})$ and $({}^s g_{\mathrm{b}}, {}^s h_{\mathrm{b}})$ for each wavelet packet tree; dashed lines show the mirrored pairs of filters $({}^s g_{\mathrm{b}}, {}^s h_{\mathrm{b}})$ for WPT A and $({}^s g_{\mathrm{a}}, {}^s h_{\mathrm{a}})$ for WPT B; dotted lines represent the filter pairs $({}^s g_{\mathrm{a}}, {}^s h_{\mathrm{a}})$ that are applied in both WPTs. For simplicity, the even and odd downsamplers are not illustrated.

This architecture is adopted from the design in [WBK09] and is almost similar to the previous structure. The same filter types and downsamplers are employed in this new architecture (cf. Fig. 5.4); nevertheless, the order of the filters is different.

As illustrated in Fig. 5.4, the solid lines in branches (1.1b), (1.1a), (2.2b), (2.2a), (3.4b), and (3.4a) are the regular pairs of filters $(^s g_\mathrm{a},\ ^s h_\mathrm{a})$ and $(^s g_\mathrm{b},\ ^s h_\mathrm{b})$ that are applied in each wavelet packet tree. This guarantees the correct phase difference needed for an analytic wavelet transformation, as in DTCWT [SBK05, BaS08].

Applying these pairs of filters in all subsequent scales results in adding an additional phase difference and non-analytic coefficients. Consequently, the filter pairs depicted in dashed and dotted lines produce an analytic transformation [WBK09]. The dashed filter branches are mirrored compared to the respective solid line branches in that scale. Filters in branches (2.1b), (2.1a), (3.1b), and (3.1a) are mirrored versions of the filters in branches (2.2b), (2.2a), (3.4b), and (3.4a). Thus, instead of applying the low-pass/high-pass filter pairs of the WPT A, the high-pass/low-pass filters of the WPT B are taken into consideration. These filters add a phase difference of $\pm 1/2\omega$ [WBK09]. The dotted lines' filters are identical in both wavelet packet trees, so they do not add any phase differences. It is possible to engage either $(^s g_\mathrm{a},\ ^s h_\mathrm{a})$ or $(^s g_\mathrm{b},\ ^s h_\mathrm{b})$ filter pairs in both trees for dotted lines [WBK09]. Before further explaining how this structure precisely provides an approximately analytic representation, some necessary definitions are presented.

ACQTWP transform utilises real-valued wavelet and scaling functions, defined as:

Definition 5.5 (ACQTWP wavelet and scaling functions). Let $\psi_{\mathrm{a},2J+1}(t), \psi_{\mathrm{a},2J+3}(t)$, $\psi_{\mathrm{b},2J+1}(t)$, $\psi_{\mathrm{b},2J+3}(t)$ and $\phi_{\mathrm{a},2J}(t), \phi_{\mathrm{a},2J+2}(t)$, $\phi_{\mathrm{b},2J}(t)$, $\phi_{\mathrm{b},2J+2}(t)$ be the wavelet and scaling functions of the ACQTWP.

The wavelet and scaling functions in WPT A, $\forall n \in \mathbb{N}$ are given by

$$^{s+1}\psi_{\mathrm{a},2J+1}(t) = \sqrt{2} \sum_{n=0}^{M} {}^s f_1[n]\ ^s\phi_{\mathrm{a},2J}(2t-n),$$

$$^{s+1}\psi_{\mathrm{a},2J+3}(t) = \sqrt{2} \sum_{n=0}^{M} {}^s f_2[n]\ ^s\phi_{\mathrm{a},2J+2}(2t-n+1),$$

$$^{s+1}\phi_{\mathrm{a},2J}(t) = \sqrt{2} \sum_{n=0}^{M} {}^s f_1[n]\ ^s\phi_{\mathrm{a},2J}(2t-n),$$

$$^{s+1}\phi_{\mathrm{a},2J+2}(t) = \sqrt{2} \sum_{n=0}^{M} {}^s f_2[n]\ ^s\phi_{\mathrm{a},2J+2}(2t-n+1).$$

Parameter $J = 2j$, where $0 \le j < 2^s \cdot (s-1)$, and $s \in \mathbb{N}$ is the number of scales. Filters f_1 and f_2 are defined by

$$(f_1, f_2) = \begin{cases} (g_\mathrm{a}, h_\mathrm{a}) & j = 0 \text{ (first branch)}, \\ (h_\mathrm{b}, g_\mathrm{b}), & j = (2^s \cdot (s-1)) - 1 \text{ (last branch)}, \\ \text{Swap the filters in each} & \\ \text{branchwith respect to} & \text{rest.} \\ \text{the previous branch.} & \end{cases}$$

For WPT B, the wavelet and scaling functions are defined in the same manner. All filters are causal, so ${}^{s}h_{\mathrm{a,b}}[n] = 0$ and ${}^{s}g_{\mathrm{a,b}}[n] = 0$ for $n < 0$.

The ACQTWP transform outcomes are complex coefficients obtained when the coefficients of both wavelet packet trees are considered. Thus, first, the wavelet and scaling coefficients of the ACQTWP for the WPT A and WPT B are defined:

Definition 5.6 (ACQTWP's coefficients of WPT A). Coefficients of ACQTWP for WPT A are given by ${}^{s}C[n] = \{\, {}^{s+1}C_{2J}[n],\ {}^{s+1}C_{2J+1}[n],\ {}^{s+1}C_{2J+2}[n],\ {}^{s+1}C_{2J+3}[n]\,\}$ and obtained by

$$
{}^{s+1}C_{2J}[n] = \sum_{k=0}^{M+Len-1} {}^{s}f_1[k]\,{}^{s}C_j[2n-k],
$$
$$
{}^{s+1}C_{2J+1}[n] = \sum_{k=0}^{M+Len-1} {}^{s}f_2[k]\,{}^{s}C_j[2n-k], \tag{5.1}
$$

$$
{}^{s+1}C_{2J+2}[n] = \sum_{k=0}^{M+Len-1} {}^{s}f_1[k]\,{}^{s}C_j[2n+1-k],
$$
$$
{}^{s+1}C_{2J+3}[n] = \sum_{k=0}^{M+Len-1} {}^{s}f_2[k]\,{}^{s}C_j[2n+1-k],
$$

where $Len = length({}^{s}C_j)$, $J = 2j$, and $0 \le j < 2^s \cdot (s-1)$. f_1 and f_2 are pairs of filters obtained by:

$$
(f_1, f_2) = \begin{cases} (g_{\mathrm{a}}, h_{\mathrm{a}}) & j = 0 \text{ (first branch)}, \\ (h_{\mathrm{b}}, g_{\mathrm{b}}), & j = (2^s \cdot (s-1)) - 1 \text{ (last branch)}, \\ \text{Swap the filters in each} & \\ \text{branch with respect to} & \text{rest.} \\ \text{the previous branch.} & \end{cases}
$$

Filters ${}^{s}h_{\mathrm{a,b}}$ and ${}^{s}g_{\mathrm{a,b}}$ are high- and low-pass filters in wavelet packets A and B.

Definition 5.7 (ACQTWP's coefficients of WPT B). The wavelet and scaling coefficients of WPT B are obtained in the same way as WPT A coefficients, and these coefficients are depicted by ${}^{s}D[n] = \{\, {}^{s+1}D_{2J}[n],\ {}^{s+1}D_{2J+1}[n],\ {}^{s+1}D_{2J+2}[n],$ ${}^{s+1}D_{2J+3}[n]\}$.

Definition 5.8 (ACQTWP complex coefficients). Complex coefficients of the ACQTWP transform are denoted by ${}^{s}\mathbb{C}$ and obtained by: ${}^{s}\mathbb{C}[n] = {}^{s}C[n] + i\,{}^{s}D[n]$, where $s \in \mathbb{N}$ is the number of scales, $i^2 = -1$, and ${}^{s}C[n]$ and ${}^{s}D[n]$ are coefficients of WPT A and B.

Besides the above definition, the coefficients of the ACQTWP transform can be divided into approximation and detail coefficients.

Definition 5.9 (Approximation and detail coefficients). The coefficients of the ACQTWP transformation for WPT A in scale s are given by approximation (scaling) ${}^{s}Q_{\mathrm{A}}[n]$ and detail (wavelet) ${}^{s}R_{\mathrm{A}}[n]$ coefficients

$$
\begin{aligned}
{}^{s}Q_{\mathrm{A},l}[n] &= \{{}^{s}C_{2J}[n],\; {}^{s}C_{2J+2}[n]\}, \\
{}^{s}R_{\mathrm{A},k}[n] &= \{{}^{s}C_{2J+1}[n],\; {}^{s}C_{2J+3}[n]\},
\end{aligned}
\tag{5.2}
$$

where $l = 2j$, $k = 2j + 1$ and $0 \le j < 2^{s}$. For WPT B, the approximation and detail coefficients are obtained similarly, and denoted by ${}^{s}Q_{\mathrm{B},l}[n]$ and ${}^{s}R_{\mathrm{B},k}[n]$. The complex approximation and detail coefficients considering both wavelet packet trees are denoted by ${}^{s}Q_{\mathbb{C},l}$ and ${}^{s}R_{\mathbb{C},k}$, respectively.

To recover the original signal, the reconstruction transformation, also known as the inverse wavelet transformation, must be applied.

5.3.1.1 Inverse Analytic Complex Quad Tree Wavelet Packet Transform (IACQTWP)

Inverse Analytic Complex Quad Tree Wavelet Packet transform (IACQTWP) reassembles the original signal using the synthesis filters.

Definition 5.10 (IACQTWP synthesis filter). Synthesis low- and high-pass filters of the inverse WPT A are respectively assigned by ${}^{s}g_{\mathrm{a}}'$ and ${}^{s}h_{\mathrm{a}}'$ and obtained by [Kin01]

$$
{}^{s}g_{\mathrm{a}}'[M - 1 - n] = (-1)^{-n}\; {}^{s}g_{\mathrm{a}}[n], \qquad {}^{s}h_{\mathrm{a}}'[M - 1 - n] = (-1)^{n}\; {}^{s}h_{\mathrm{a}}[n],
$$

where $M \in \mathbb{N}$ is the length of the filters. Synthesis filters of the inverse WPT B are computed similarly.

By utilising the synthesis filters, the original data are reconstructed from the coefficients obtained in the decomposition stage.

Definition 5.11 (IACQTWP coefficients). Coefficients of the IACQTWP for inverse WPT A are reconstructed by

$$
\begin{aligned}
{}^{s}C_{j}^{\mathrm{I}}[n] = \sum_{k=0}^{M+Len-1} {}^{s}f_{1}'[k]\; {}^{s+1}C_{2J}[2n - k] + {}^{s}f_{2}'[k]\; {}^{s+1}C_{2J+1}[2n - k] + \dots \\
+\; {}^{s}f_{1}'[k]\; {}^{s+1}C_{2J+2}[2n - k] + {}^{s}f_{2}'[k]\; {}^{s+1}C_{2J+3}[2n - k],
\end{aligned}
$$

where f_{1}' and f_{2}' are obtained by:

$$
(f_{1}', f_{2}') = \begin{cases}
(g_{\mathrm{a}}', h_{\mathrm{a}}') & j = 0 \text{ (first branch)}, \\
(h_{\mathrm{b}}', g_{\mathrm{b}}'), & j = (2^{s} \cdot (s - 1)) - 1 \text{ (last branch)}, \\
\text{Swap the filters in each} & \\
\text{branch with respect to} & \text{rest.} \\
\text{the previous branch.} &
\end{cases}
$$

Low- and high-pass synthesis filters in the inverse wavelet packet trees A and B are depicted by $g'_{a,b}$ and $h'_{a,b}$. Coefficients of inverse WPT B, $^sD_j^I[n]$, are obtained likewise.

After introducing the ACQTWP transform, its main characteristics must be investigated. Like other wavelet transformations, the ACQTWP transformation has properties such as efficient implementation (filter bank implementation) and multi-resolution analysis. Other properties of the ACQTWP transform are highlighted in the succeeding sections.

5.3.1.2 Properties and Characteristics

In addition to the aforementioned properties, the ACQTWP transform provides an approximate analytic representation of the signal and is shift invariant. These latest properties of the ACQTWP transform are described in this section.

Analytic Representation

The amplitude spectra of the proposed analytic wavelet transform (ACQTWP) and the non-analytic one are illustrated in Fig. 5.5. As depicted, the first scale in both transformations is not analytic, since the filters do not follow the half-sample delay condition necessary for constructing the Hilbert pair wavelets (refer to Sec. 2.4.1 for more information). Contrary to the non-analytic transformation, for ACQTWP transform, the frequency response of the scales $s > 1$ is close to one-sided.

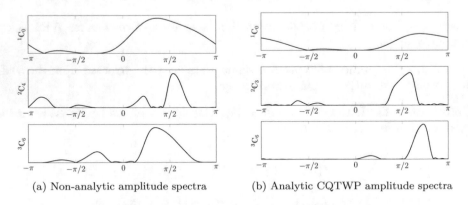

(a) Non-analytic amplitude spectra   (b) Analytic CQTWP amplitude spectra

Figure 5.5: The amplitude spectra of basis functions in three scales, when applying the proposed non-analytic transform (a) and ACQTWP transform (b). The outputs of three branches are given. These are the complex coefficients obtained by applying the coefficients in both wavelet packet trees.

However, ACQTWP is approximately analytic, which means that some amount of energy exists in the negative frequency band, as presented in Fig. 5.5 (b) for $^3\mathbb{C}_3$.

The amplitude spectrum of $^3C_3$ is approximately analytic, and some amount of energy leeks in the negative frequency band. This issue is not due to the ACQTWP transform structure but to the Hilbert transform [Ciz70]. The Hilbert transform extends infinitely in both the time and frequency domain. This means that the Hilbert transform of a compactly-supported wavelet or a scaling function cannot be of finite support [Aus93,SBK05]. Thus, any precisely analytic Hilbert pair of wavelet functions must have infinite support. On the other hand, the half-sample delay condition can only be approximated with finite length filters [Sel01]. Therefore, the wavelet functions of the ACQTWP must be designed with FIR filters [OpS89], which results in wavelet functions with finite support, which indicates that wavelets of the ACQTWP are approximately analytic.

In order to explain how the ACQTWP structure satisfies the half-sample delay condition, the reconstruction transforms, as illustrated in Fig. 5.6, must be considered.

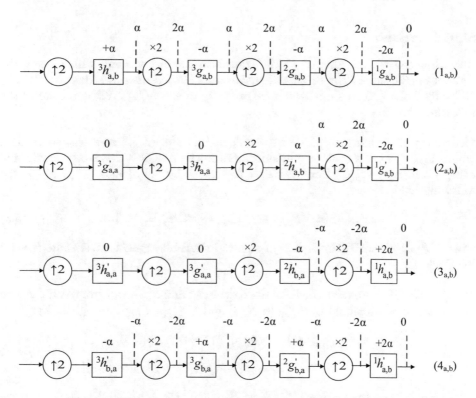

Figure 5.6: Phase difference of each complex sub-tree; filters $^sh'_{a,b}$ and $^sg'_{a,b}$ are the synthesis high- and low-pass filters. Index (a,b) depicts the order of filters in the regarded trees so that $^sg'_{a,b}$ are the synthesis low-pass filters of tree A and tree B with $-\alpha$ phase difference, where $^sg'_{b,a}$ are the synthesis low-pass filters of tree B and A with α phase difference. $\alpha = 1/2\omega$. $\uparrow 2$ is the upsampling operator.

Figure 5.6 represents the phase difference in each of the complex sub-trees, considering the inverse ACQTWP transform. The input for the inverse tree is the output of each analysis wavelet packet tree in Fig. 5.4. The filters depicted in Fig. 5.6 are the synthesis filters of both wavelet packet trees. The filter index depicts their order concerning the applied tree so that $^1g'_{a,b}$ are the synthesis low-pass filters of the first scale in tree A and tree B, where $^1g'_{b,a}$ are the synthesis low-pass filters of the first scale in tree B and A. The phase difference contributed by each filter in each scale is depicted in Fig. 5.6, where $\alpha = 1/2\omega$. The synthesis high-pass filters $^sh'_{a,b}$ and the synthesis low-pass filters $^sg'_{a,b}$ add phase differences of $\pm\alpha$. The synthesis low- and high-pass filters $^sg'_{a,a}$ and $^sh'_{a,a}$ do not contribute any phase difference because similar filters are applied in both trees. Filters of the first scale donate $\pm2\alpha$ since they have a phase difference of one sample delay. By computing phase differences in each sub-tree, it is indicated that the reconstructed signal has no phase difference with the input signal.

Shift Invariance

According to Def. 2.17, the ACQTWP transform is shift invariant, which means the wavelet coefficients of the original signal and its shifted ones are identical [Kin00, GGS$^+$09, ToL15]. The shift-invariance property of the ACQTWP is proved in the following corollary.

Corollary 2. Assume $x[n]$ is a discrete signal and $x[n - S_{e/o}]$ is its shifted version, where $S_{e/o} \in \mathbb{Z}$ takes even or odd shifts. Given the ACQTWP wavelet coefficients of $x[n]$ and $x[n - S_{e/o}]$ from WPT A in scale s by $^sC[n]$ and $^sC'[n, S_{e/o}]$, the following holds:

$$\forall x[n],\ x[n - S_{e/o}] \in L^2(\mathbb{R}) \ \Rightarrow\ ^sC[n] \cong\ ^sC'[n], \tag{5.3}$$

where $L^2(\mathbb{R})$ is the Hilbert space of square integrable signals. The same is valid for WPT B.

Proof. ACQTWP wavelet coefficients of $x[n - S_e]$ and $x[n - S_o]$ from WPT A and B in scales s are depicted by $^sC'_{e/o}[n, S_{e/o}]$ and $^sD'_{e/o}[n, S_{e/o}]$, and given by:

$$^sC'_{e/o}[n, S_{e/o}] = \{^{s+1}C'_{2J}[n, S_{e/o}],\ ^{s+1}C'_{2J+1}[n, S_{e/o}], \cdots$$
$$\cdots\ ^{s+1}C'_{2J+2}[n, S_{e/o}],\ ^{s+1}C'_{2J+3}[n, S_{e/o}]\},$$

$$^sD'_{e/o}[n, S_{e/o}] = \{^{s+1}D'_{2J}[n, S_{e/o}],\ ^{s+1}D'_{2J+1}[n, S_{e/o}], \cdots$$
$$\cdots\ ^{s+1}D'_{2J+2}[n, S_{e/o}],\ ^{s+1}D'_{2J+3}[n, S_{e/o}]\}.$$

To provide a shift-invariant property, a non-shifted and a shifted version of the input signal in each scale are decomposed. The coefficients of signal $x[n - S_{e/o}]$ for both odd and even shifts are given in the following [ToL15]:

1. **Even Shifts.** If $x[n - S_e]$, where $S_e = 2m$, $m \in \mathbb{Z}$, then ACQTWP's

coefficients $^{s+1}C'_{2J}[n, S_e]$ and $^{s+1}C'_{2J+1}[n, S_e]$ are equivalent to $^{s+1}C_{2J+1}[n]$. Thus,

$$^{s+1}C'_{2J}[n, S_e] \overset{S_e=2m}{=} \sum_{k=0}^{M+Len-1} {}^sg_a[k] \, {}^sC'_j[2n - 2m - k] = \ldots$$

$$\ldots = \sum_{k=0}^{M+Len-1} {}^sg_a[k] \, {}^sC'_j[2(n-m) - k] = {}^{s+1}C_{2J}[n-m],$$

$$^{s+1}C'_{2J+1}[n, S_e] \overset{S_e=2m}{=} \sum_{k=0}^{M+Len-1} {}^sh_a[k] \, {}^sC'_j[2n - 2m - k] = \ldots \tag{5.4}$$

$$\ldots = \sum_{k=0}^{M+Len-1} {}^sh_a[k] \, {}^sC'_j[2(n-m) - k] = {}^{s+1}C_{2J+1}[n-m].$$

2. Odd Shifts. If $x[n-S_o]$ with shift $S_o = 2m+1, \quad m \in \mathbb{Z}$, then ACQTWP's coefficients $^{s+1}C'_{2J+2}[n, S_o]$ and $^{s+1}C'_{2J+3}[n, S_o]$ are equivalent to $^{s+1}C_{2J+1}[n]$. Consequently,

$$^{s+1}C'_{2J+2}[n, S_o] \overset{S_o=2m+1}{=} \sum_{k=0}^{M+Len-1} {}^sg_a[k] \, {}^sC'_j[2n + 1 - 2m - 1 - k] = \ldots$$

$$\ldots = \sum_{k=0}^{M+Len-1} {}^sg_a[k] \, {}^sC'_j[2(n-m) - k] = {}^{s+1}C_{2J}[n-m],$$

$$^{s+1}C'_{2J+3}[n, S_o] \overset{S_o=2m+1}{=} \sum_{k=0}^{M+Len-1} {}^sh_a[k] \, {}^sC'_j[2n + 1 - 2m - 1 - k] = \ldots \tag{5.5}$$

$$\ldots = \sum_{k=0}^{M+Len-1} {}^sh_a[k] \, {}^sC'_j[2(n-m) - k] = {}^{s+1}C_{2J+1}[n-m].$$

Similarly, the coefficients for the second wavelet packet, WPT B, are obtained. So,

$$\begin{cases} \forall \, x[n - S_e], & \begin{cases} {}^sC[n] = {}^sC'_e[n - \lceil \frac{S_e}{2^s} \rceil], \\ {}^sD[n] = {}^sD'_e[n - \lceil \frac{S_e}{2^s} \rceil]. \end{cases} \\[2ex] \forall \, x[n - S_o], & \begin{cases} {}^sC[n] = {}^sC'_o[n - \lceil \frac{S_o}{2^s} \rceil], \\ {}^sD[n] = {}^sD'_o[n - \lceil \frac{S_o}{2^s} \rceil]. \end{cases} \end{cases} \tag{5.6}$$

Based on Eq. 5.6, for $x[n]$ and $x[n - S_{e/o}]$, $^sC[n] = {}^sC'[n - \kappa]$, where $\kappa = \lceil \frac{S_{e/o}}{2^s} \rceil$. Accordingly, $^sC[n] \simeq {}^sC'[n]$ and the ACQTWP transform is said to be shift invariant. $\qquad\square$

As proved, the ACQTWP transform is shift invariant. Thus, based on Def. 2.17, the ACQTWP mapping builds equivalence classes, where all similar time series belong to one class. Nevertheless, only equivalence relations produce equivalence classes.

For this reason, it must be shown that besides being a relation, the ACQTWP transform is also an equivalence relation.

Lemma 3. The ACQTWP transform is an equivalence relation.

Proof. Let $x_1 = x[n]$, $x_2 = x[n - p]$ and $x_3 = x[n - q]$, where $p, q = S_{e/o} \in \mathbb{Z}$. Based on the definition of equivalence relations, the ACQTWP transformation is an equivalence relation, since:

- Reflective: $\forall x_1 \in L^2(\mathbb{R})$, $(x_1, x_1) \in$ ACQTWP, so ACQTWP$(x_1) \cong$ ACQTWP(x_1).

- Symmetric: $\forall x_1,\ x_2 \in L^2(\mathbb{R})$, if $(x_1, x_2) \in$ ACQTWP, then as ACQTWP is shift-invariant (Lemma 2) ACQTWP$(x_2) \cong$ ACQTWP(x_1), and so $(x_2, x_1) \in$ ACQTWP.

- Transitive: $\forall x_1,\ x_2,\ x_3 \in L^2(\mathbb{R})$, if $(x_1, x_2) \in$ ACQTWP and $(x_2, x_3) \in$ ACQTWP, then based on the Lemma 2 ACQTWP$(x_1) \cong$ ACQTWP(x_3), so $(x_1, x_3) \in$ ACQTWP.

□

Based on the lemma 3, the equivalence relation ACQTWP builds the following equivalence class

$$[x] = \{\forall\ x_i[n] \in L^2(\mathbb{R}),\ i \in \mathbb{N} | (x_i, x) \in \text{ACQTWP}\}.$$

The three signals x_1, x_2, and x_3 in Lemma 3 are members of the equivalence class $[x]$.

Similarly, ACQTWP builds two equivalence classes $[Q]$ and $[R]$ on the set of all the decomposed coefficients, where class $[Q]$ contains all the equivalence approximation coefficients, and class $[R]$ includes all the equivalence detail coefficients.

ACQTWP's shift-invariant property is provided in the following examples.

Example 3. Let $x[n]$ be a step signal which, along with its shifted versions, is given in Fig. 5.7 (a). Sub-figures (b)-(d) represent the ACQTWP coefficients obtained within three scales. As depicted, in each scale, all of the wavelet coefficients are identical.

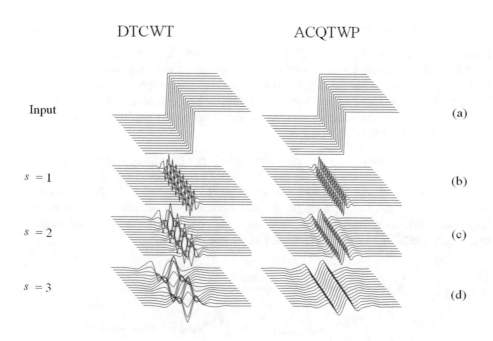

Figure 5.7: Effect of shifts in three scales of the DTCWT and ACQTWP transform. (a) a step signal and its shifted versions are illustrated. (b)-(d)
coefficients of DTCWT and ACQTWP transform within three scales
are depicted.

5.3.1.3 Selection of the Best Basis

While examining a signal with the ACQTWP transform, it is unnecessary to consider all the frequency scales that the signal is decomposed into due to information
redundancy. This can also become a time-consuming task. Consequently, it is common to define the "Best Basis" [Wic96, BaS08, GGS⁺09, WBK09, TDA15, CBA16,
BPC⁺18, WRH18].

Definition 5.12 (Best Bases). Best Bases are specific wavelet bases that provide
the most desirable representation of the signal [Wic96].

If the bases are orthogonal, they are able to approximate signals with only a few
non-zero vectors. Thus, by selecting the wavelet packet basis that concentrates the
signal energy over a few coefficients, its time-frequency structure is also ascertained
[BaS08, WBK09]. In this section, a "Best Basis" algorithm is proposed, which
follows a criterion for stopping the wavelet decomposition and selecting nodes with
the most extensive information content.

Early Abandoning

Instead of analysing the ACQTWP transform into the total amount of scales, $s_T \in \mathbb{N}$, and analysing them, a method is presented to decompose a signal up to a specific scale. This scale must contain the most extensive information content and also preserve the shape of the signal. By increasing the scales of the ACQTWP transform from coarse to fine, the input signal loses its original shape. This is illustrated in the following example.

Example 4. Subsequence m_1 from Example 2 is decomposed to eight scales by the ACQTWP transform. Fig. 5.8 (a) represents subsequence m_1 and the WPT A approximation coefficients of four scales: $^1C_0$, $^3C_0$, $^5C_0$, and $^7C_0$ (cf. sub-figures (b) and (c)). Although the length of these coefficients is less than the original signal m_1, due to downsampling, $^1C_0$ and $^3C_0$ coefficients preserve the shape of m_1. Coefficients of $^5C_0$ and $^7C_0$ do not resemble signal m_1.

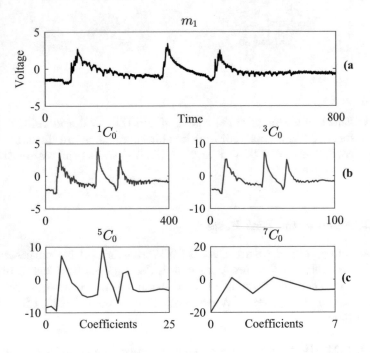

Figure 5.8: Decomposition of signal m_1 by the ACQTWP transform up to eight scales. The approximation coefficients of four scales are presented (b)-(c). The approximation coefficients $^5C_0$ and $^7C_0$ do not preserve the original shape of the signal m_1.

The shape of a signal's distribution is described by shape factors. The third and the fourth standard moment, skewness and kurtosis [BaM88, BaM88], are the most well-known shape factors. Skewness is a measure of symmetry [Pea05] (as given in [BaM88]). A negative skewness illustrates that the distribution is skewed to the left, and the distribution's tail is longer on the left. On the other hand, a positive skewness means that the distribution is skewed to the right [BaM88].

Kurtosis is a descriptor of the shape of a probability distribution, and it is a measure of the tails of the distribution [BaM88]. High kurtosis implies that outliers are present in the distribution.

The concept of shape factors is employed to measure the similarity between the shape of the signals (input and wavelet coefficients) and stop the ACQTWP decomposition (cf. 2.7). As skewness of the wavelet coefficients provides a degree of symmetry, its information gain is not enough to compare the coefficients' shape with the original signal's shape. Therefore, kurtosis is employed to compare the shape of the wavelet coefficients of ACQTWP with the original signal.

Definition 5.13 (ACQTWP shape factor). ACQTWP shape factor of WPT A is denoted by $^s\beta_{A,l}$ and is defined by [BaM88]

$$^s\beta_{A,l} = \left(\frac{E[(\ H(^sQ_{A,l}) - E[H(^sQ_{A,l})])^4]}{(E[(\ H(^sQ_{A,l}) - E[H(^sQ_{A,l})])^2])^2} \right),$$

where $l = 2j$, $0 \leq j < 2^s$ and $1 \leq s \leq s_T$. In each scale, $^s\beta_{A,l}$ measures the kurtosis of the histograms of the approximation coefficients of WPT A $(H(^sQ_{A,l}))$. Similarly, ACQTWP shape factor of WPT B, $^s\beta_{B,l}$, is obtained.

Note: It should be noted that the definition of kurtosis is different from the excess kurtosis [DeC97], which is the kurtosis minus three. The excess kurtosis only provides a comparison to the normal distribution.

In each scale, the shape of the approximation coefficients is compared with the original signal's shape. This is performed by computing the minimum difference between the ACQTWP shape factor of the approximation coefficients and the input signal. So,

$$^s\Delta = \min_{l=0}^{L}(\ ^s\Delta_{A,l}, \ ^s\Delta_{B,l}), \tag{5.7}$$

where

$$^s\Delta_{A,l} = |1 - \frac{^s\beta_{A,l}}{\beta_x}|, \qquad ^s\Delta_{B,l} = |1 - \frac{^s\beta_{B,l}}{\beta_x}|.$$

The $^s\Delta_{A,B}$ is rescaled to be bounded between $0 \leq^s \Delta_{A,B} < 1$. The closer $^s\Delta_{A,B}$ is to zero, the more similar the two signals are to each other. By considering Def. 2.7 and the definition of the shape factor, the following equivalence class is obtained.

Definition 5.14 (Class of approximation coefficients similar to $x[n]$). Assume the set S_C comprises signal $x[n]$ and all approximation coefficients obtained from both WPT A and B. The equivalence relation R_k builds an equivalence class C_E on the set $S_C = \{\{^sQ_{A,B}[n]\} \cup \{x[n]\}\}$, $S_C \subset L^2(\mathbb{R})$. So,

$$[C_E] = \{^sQ_l \in S_C | (\ ^sQ_l, x) \in R_k\},$$

where $^sQ_l \in \{^sQ_{A,B}[n]\}$ and s and $l \in \mathbb{N}$ are the scale and the index of the coefficients' node.

The equivalence class $[C_E]$ contains all approximation coefficients similar to the original signal $x[n]$.
The concept of a shape factor is employed to abort the decomposition when the shape of the wavelet coefficients is not similar to the original signal. After stopping the decomposition, the scale with the highest amount of information must be selected. In the information theory, Shannon defined the amount of information by means of entropy [Sha01]. Entropy is also a measure of outliers and uncertainty in the data. The wavelet entropy of the ACQTWP wavelet coefficients is measured as follows:

Definition 5.15 (Wavelet entropy [Wic96]). The entropy for complex wavelet coefficients of the ACQTWP transform is denoted by $^sEn(\mathbb{C}_i)$ and computed by [Wic96]

$$^sEn(\mathbb{C}_i) = -\sum_{n=1}^{N} E_c[n] \log(E_c[n]),$$

where $E_c[n] = \frac{(^s\mathbb{C}[n])^2}{\sum_n (^s\mathbb{C}[n])^2}$ is the normalised energy of the complex wavelet coefficients, so $\sum_{n=1}^{N} E_c[n] = 1$, and if $E_c[i] = 0$, then $E_c[i]\log(E_c[i]) = 0$ [Wic96]. Index i refers to the nodes in scale s, so that $0 \leq i < 4^s$, and $1 \leq s \leq s_T$.

Besides entropy, uncertainty is also estimated by variance [GaM56]. Variance and the energy of a discrete signal are computed similarly. In contrast to the variance equation [PoM07], energy is determined without removing the arithmetic mean of the signal.
Thus, to detect the scale with the highest amount of information and the minimum amount of outliers, energy and entropy are considered. In the wavelet theory, analysing and comparing the amount of energy in various scales is only meaningful if the wavelet functions are orthogonal, and this is based on the Parseval's theorem [BGG+98].
As shown in Appendix 9.2, the filters of the ACQTWP transform are orthonormal. Thus, for these wavelets, the Parseval's theorem says the following:

Theorem 3 (Parseval's theorem for wavelets). Iff the wavelet functions are orthonormal, then Parseval's theorem (original paper in French [Par06]) [BGG+98] states that the energy of the signal $x[n]$ and each scale's coefficients are proportional to each other.

Proof. The proof is given in Appendix 10.1. $\square$

According to Theorem 3 and the ACQTWP transform filters' design (cf. Appendix 9.2), the amount of energy is preserved in the ACQTWP transform. Consequently, the energy of the signal is equal to the energy of the scaling (approximation) and wavelet (detail) coefficients [BGG$^+$98]:

$$\sum_{n}(|Q_A[n]|^2 + |R_A[n]|^2) = \sum_{n}|x[n]|^2,$$

where $Q_A[n]$ and $R_A[n]$ are approximation and detail coefficients of WPT A. The same applies to WPT B. This indicates that the amount of signal's energy is separated by means of the scales of this wavelet transformation.

Due to downsampling, the amount of energy of different scales cannot be compared directly, so the energy density of the coefficients must be obtained in order to examine the amount of energy of different scales.

Definition 5.16 (Energy density). The energy density of the complex coefficients of both wavelet packet trees A and B, $^sQ_C[n] = \, ^sQ_A[n] + i \, ^sQ_B[n]$, in each scale is obtained as

$$^sE_D(Q_{C,l}) = \frac{\sum_{n}(|^sQ_{C,l}[n]|)^2}{N}, \qquad (5.8)$$

where $^sQ_{C,l}[n] = (C_1, C_2, ..., C_N)$ and $N \in \mathbb{N}$ is the length of the coefficients. Parameter s and the index l are respectively defined by $1 \le s \le s_T$, and $l = 2j$, $0 \le j < 2^s$. Equation 5.8 provides the relative proportion of energy to the length of the coefficients in each scale.

The best scale must be selected so that the energy concentration is the highest and entropy has its minimum value. A large value of energy density hints at the absence of outliers and concentrated data. A lower amount of entropy shows higher energy concentration. Consequently, a measure based on both concept of energy density and entropy, namely the energy-to-entropy ratio $^s\eta(C_i)$, is proposed.

Definition 5.17 (Energy-to-entropy ratio). The energy to entropy ratio is denoted by $^s\eta$ and is obtained by

$$^s\eta(C_i) = \frac{^sE_D(C_i)}{^sEn(C_i)}, \qquad (5.9)$$

where $^sE_D(C_i)$ is the energy density of the complex wavelet coefficients, (cf. Eq. (5.8)), and $^sEn(C_i)$, (cf. Def. 5.15), is the entropy of the complex wavelet coefficients. The nodes in scale s are presented by the index i, $0 \le i < 4^s$.

It should be noted that only the approximation coefficients are considered to abort the decomposition since, based on the definition of wavelet transformations, the signal's shape is mostly conserved in the approximation coefficients. Nevertheless, after stopping the decompositions, the energy-to-entropy ratio of both detail and approximation coefficients is analysed in order to detect the most proper scale $\theta = s_b$.

Early abandoning algorithm applies both energy-to-entropy ratio and the shape factor concepts in its steps:

1. (Stopping the decomposition): The difference between the kurtosis of the wavelet coefficients and the original signal is measured. If this amount is

less than the threshold τ_Δ, so that $^s\Delta \leq \tau_\Delta$, then $^sQ_l[n] \in [C_{\mathrm{E}}]$, and the decomposition is carried on. Otherwise, the decomposition is discontinued. After aborting the decomposition, the best scale, which is denoted by s_{b}, is selected.

2. (Selecting the best scale): If the amount of the energy-to-entropy ratio of the complex approximation $(\eta(Q_{\mathbb{C},l}))$ and detail $(\eta(R_{\mathbb{C},l}))$ coefficients drops, so that

$$\left| \frac{^{s-1}\eta(Q_{\mathbb{C},l}) - \,^s\eta(Q_{\mathbb{C},l})}{^{s-1}\eta(Q_{\mathbb{C},l})} \right| > \left| \frac{^s\eta(Q_{\mathbb{C},l}) - \,^{s+1}\eta(Q_{\mathbb{C},l})}{^s\eta(Q_{\mathbb{C},l})} \right|$$

and

$$\left| \frac{^{s-1}\eta(R_{\mathbb{C},l}) - \,^s\eta(R_{\mathbb{C},l})}{^{s-1}\eta(R_{\mathbb{C},l})} \right| > \left| \frac{^s\eta(R_{\mathbb{C},l}) - \,^{s+1}\eta(R_{\mathbb{C},l})}{^s\eta(R_{\mathbb{C},l})} \right|$$

are fulfilled, then s is the scale with the most information content. If the value of the energy-to-entropy ratio increases, then the scale with the maximum amount of this ratio is considered as the best scale.

The threshold $0 \leq \tau_\Delta < 1$ provides an upper limit to decides on the dissimilarity (shape difference) between the original signal and its wavelet coefficients and it must be provided in advance. As motif discovery aims to detect similar subsequences, smaller values should be assigned to τ_Δ. For the rest of this work, τ_Δ is considered to be $0 \leq \tau_\Delta < 0.5$. As an example, $\tau_\Delta = 0.3$ means that the decomposition is terminated when the shape difference between the approximation coefficients and the original signal is more than $30\,\%$. Assigning larger values to τ_Δ ($\tau_\Delta \geq 0.5$) results in the further decomposition of the wavelet tree and increasing the amount of redundant information and the time complexity of the ACQTWP transform. The procedure of the early abandoning method is described in Algorithm 3.

Lemma 4. Algorithm 3 stops the ACQTWP decomposition before the s_{T}-th scale.

Proof. According to the early abandoning algorithm, the ACQTWP decomposition is terminated based on the shape difference condition.
By increasing the scales, at some point, the length of the coefficients is so small (e.g., four or eight samples) that the shape of the original signal is not recognizable. Thus, $\exists\, s_m$, $1 < s_m < s_{\mathrm{T}}$, so that $^{s_m}\Delta \leq \tau_\Delta$, and therefore, the coefficients $^{s_m}Q_l[n]$ do not belong to the equivalence class $[C_{\mathrm{E}}]$, $^{s_m}Q_l[n] \not\approx x[n] \Rightarrow \,^{s_m}Q_l[n] \notin [C_{\mathrm{E}}]$, which results in stopping the decomposition at scale s_m. $\qquad\square$

The next example illustrates the performance of the early abandoning algorithm.

Example 5. By executing Algorithm 3 for signal m_1 in Example 4, the decomposition is interrupted after the third scale, as presented in Table 5.1. By setting $\tau_\Delta = 0.2$, the wavelet decomposition after the fourth scale discontinues. This means the coefficients' shape is preserved up to the fourth scale since $^4\Delta > \tau_\Delta$. Next, by dropping the amount of energy-to-entropy ratio at the second scale, $s_{\mathrm{b}} = 2$ is selected.

Algorithm 3 Early abandoning

Input: signal or time series $x[n]$, approximation coefficients of each scale ${}^sQ_{A,l}[n]$
and ${}^sQ_{B,l}[n]$, threshold τ_Δ

Output: selected scale s_b

1: **while** $s \le s_T$ **do**
2:      **if** ${}^s\Delta \le \tau_\Delta$ **then**
3:          continue the decomposition;
4:      **else**
5:          stop the decomposition;
6:          $s = s_{\text{Stop}}$.
7:      **end if**
8: **end while**
9: **for** $i = 1 : s_{\text{Stop}}$ **do**
10:      **if** $\left| \frac{{}^{i-1}\eta(Q_{C,l}) - {}^i\eta(Q_{C,l})}{{}^{i-1}\eta(Q_{C,l})} \right| > \left| \frac{{}^i\eta(Q_{C,l}) - {}^{i+1}\eta(Q_{C,l})}{{}^i\eta(Q_{C,l})} \right|$ and
11:      $\left| \frac{{}^{s-1}\eta(R_{C,l}) - {}^s\eta(R_{C,l})}{{}^{s-1}\eta(R_{C,l})} \right| > \left| \frac{{}^s\eta(R_{C,l}) - {}^{s+1}\eta(R_{C,l})}{{}^s\eta(R_{C,l})} \right|$ **then**
12:          $i = s_b$ is the scale with the high information content.
13:      **else**
14:          find i so that $({}^i\eta(Q_{C,l}))$ has the maximum value;
15:          $i = s_b$.
16:      **end if**
17: **end for**

As proved in Sec. 5.3.1.2, the ACQTWP transform is shift invariant. Hence, the best scale s_b must be equal for the original signal and its shifted versions. In the next proposition, the effects of shift on the early abandoning algorithm are investigated.

Proposition 1. For both signal $x[n]$ and its shifted versions $x[n - S]$, $S \in \mathbb{Z}$, the decomposition of ACQTWP transform stops at the same scale, and the best scale detected by Algorithm 3 does not vary, even if the signal is shifted.

Proof. Based on Corollary 2, ACQTWP's wavelet coefficients of the original signal and its shifted ones are similar to each other, so

$$x[n] \cong x[n - S_e] \Rightarrow {}^sQ_{C,l}[n] \cong {}^s Q'_{C,l}[n - \lceil \frac{S_e}{2^s} \rceil],$$

$$x[n] \cong x[n - S_o] \Rightarrow {}^sQ_{C,l}[n] \cong {}^s Q'_{C,l}[n - \lceil \frac{S_o}{2^s} \rceil],$$

(5.10)

where ${}^sQ_{C,l}$ are the complex approximation coefficients of the ACQTWP transform, $l \in \mathbb{N}$ is the index, and $S_{e/o} \in \mathbb{Z}$ are the even and odd shift. The same holds for the detail coefficients ${}^sR_{C,k}$.

Considering Eq. 5.10 and according to Def. 5.16, the energy density and entropy of the approximation coefficients of the signal and its shifted versions are similar. Accordingly, their energy-to-entropy ratio is the same. Moreover, the shape

Table 5.1: The energy-to-entropy ratio $^s\eta$ and the shape differences $^s\Delta$ for subsequence $m_1[n]$ (cf. Example 4): the decomposition stops after the third scale since $^4\Delta > \tau_\Delta$. The most informative scale is $s_b = 2$ since the difference between the energy-to-entropy ratio of the two consequent scales drops.

Scale	$^s\Delta$	$(\,^s\eta - \,^{s+1}\eta)/\,^s\eta$
1	0.003	0.491
2	0.016	0.492
3	0.036	0.481
4	0.247	0.434
5	0.383	0.390
6	0.518	0.371

difference (Def. 5.13) for these signals is also the same, as kurtosis is shift invariant [BaM88, DeC97]. Subsequently,

$$^s\eta(Q_{C,l}) \cong \,^s\eta'(Q'_{C,l}) \quad \text{and} \quad ^s\beta \cong \,^s\beta'.$$

This indicates that the selected scale by early abandoning algorithm is similar for signal $x[n]$ and signal $x[n - S]$, regardless of the shift distortion.

The same holds for the detail coefficients.   □

Proposition 1 results in the equivalence class $[C'_E]$ under the equivalence relation R_k. So, $[C'_E] = [C_E] \cup [x]$, where $[x] = \{x_i \in L^2(\mathbb{R})|(x_i, x) \in \text{ACQTWP}\}$. Besides, the original signal $x[n]$ and its shifted versions $x[n - S]$, the equivalence class $[C'_E]$ contains the approximation coefficients $^sQ_{C,l}$ and the shifted coefficients $^sQ'_{C,l}$. The following example illustrates the effects of time-shifting in the early abandoning algorithm.

Example 6. Consider signal $m_1[n]$ from Example 2 and two shifted versions of it, $m_2 = m_1[n + 2548]$ and $m_5 = m_1[n - 1027]$. Based on Proposition 1, the outcome of Algorithm 3 is similar for all three subsequences. This is depicted in Table 5.2, where, based on the factor's concept and given the threshold $\tau_\Delta = 0.2$, the decomposition aborts after the third scale. Subsequently, as the amount of energy-to-entropy ratio $^s\eta$ drops at the second scale for all three subsequences, Algorithm 3 considers the second scale as the scale with the maximum amount of information, $s_b = 2$, for all these signals, regardless of their time translation.

After interrupting the decomposition and selecting the best scale, the most proper nodes with the highest amount of information are considered. The next section describes an approach for determining the most advantageous nodes.

Table 5.2: Energy density E_D and the shape difference Δ in five scales for three motifs of Example 6, $m_1[n]$, $m_2 = m_1[n + 2548]$, and $m_3 = m_1[n - 1027]$ are given. Based on algorithm 3 for all these subsequences, the best scale is the second scale, regardless of signal shifts.

Scale	$({}^s\eta - {}^{s+1}\eta)/\,{}^s\eta$			Δ		
	m_1	m_2	m_5	m_1	m_2	m_5
1	0.491	0.491	0.491	0.003	0.003	0.003
2	0.492	0.492	0.492	0.016	0.017	0.017
3	0.480	0.481	0.481	0.036	0.066	0.089
4	0.432	0.434	0.434	0.247	0.262	0.335

Selecting Best Nodes

With each scale of the ACQTWP transform, the number of nodes proliferates. Consequently, choosing the most informative nodes from the ACQTWP transform decreases redundant and unnecessary information. It also reduces time complexity and increases the efficiency of the algorithm [Wic96, BaS08, GGS+09, HTW15, WRH18].

The "best complex approximation and detail nodes" are defined in:

Definition 5.18 (Best Complex Approximation and Detail Nodes). Let $X = \{0, 1, ..., 4^{s_b}\}$, $Y = \{{}^{s_b}\eta_{Q0}, {}^{s_b}\eta_{Q1}, ..., {}^{s_b}\eta_{Q4^{s_b}}\}$, and Z be the totally ordered set of Y. Then,

$$^sBQ_{\mathbb{C},i} = \arg\max_i f(i),$$

where $^sBQ_{\mathbb{C},i}$ is the best complex approximation node, and argument of the maxima for $f : X \to Z$ is given by $\arg\max_i f(i) =: \{i | \forall j : f(j) \leq f(i)\}$. Similarly, the best complex detail node is computed and denoted $^sBR_{\mathbb{C},i}$.

The ability to exclude nodes with unnecessary information and to select the most informative ones relies on the energy-to-entropy ratio $^s\eta(\mathbb{C}_i)$ (Def. 5.17), introduced in the previous section. This measure aims to locate a node with the maximum amount of energy density $E_D(\mathbb{C}_i)$ and the minimum amount of entropy $En(\mathbb{C}_i)$. The best complex approximation and detail nodes are obtained by computing the ratio $^s\eta$, as given in Def. 5.18. These nodes are determined by the Best Node selection (BNS) algorithm given in Algorithm 4.

It should be noted that in contrast to most of the best base selection methods [CBA16, RMB16, WRH18], which choose the most advantageous nodes in each scale, the BNS algorithm detects the two conforming nodes (one approximation and one detail node) in the best-selected scale s_b. The reason is that the early abandoning (Algorithm 3) outcome for signals with various properties (e.g. sampling frequency, length) is diverse, which means that dissimilar signals are decomposed into a varying number of scales. Hence, selecting the most proper node from each scale (from s_b up to the first scale) leads to a different number of nodes. As the best nodes are forwarded to the feature extraction step, then the number of signals' features would not be the same, making the direct comparison of signals burdensome.

Algorithm 4 Best Nodes Selection (BNS)

Input: approximation and detail coefficients $^{sb}Q_i$ and $^{sb}R_i$

Output: best approximation and detail nodes

1: **for** $i = 1 : 1 : 4^{sb}$ **do**
2:      $^{sb}\eta_{Qi} = \frac{^{sb}ED(Q_i)}{^{sb}En(Q_i)}$
3:      $^{sb}\eta_{Ri} = \frac{^{sb}ED(R_i)}{^{sb}En(R_i)}$
4: **end for**
5: $^{s}BQ_{\mathbb{C},i} = \arg\max(^{sb}\eta_{Qi})$, and $^{s}BR_{\mathbb{C},i} = \arg\max(^{sb}\eta_{Ri})$

Next, the impact of translation mappings on the BNS algorithm is explained.

Effects of Shift on Selecting the Best Nodes

Due to ACQTWP's shift-invariant property, the best-selected scale s_b for both an original signal and its shifted ones is the same. In this section, the outcomes of the BNS algorithm concerning time translations are examined. However, first, the following information is required.

Definition 5.19 (Modulo equivalence for 2^a). Assume R is an equivalence relation on the set $S_e = \{2, 4, 6, 8, ...\}$, defined by pRq iff $\forall \, p, q \in S_e \, p \equiv q(mod \, 2^a)$. Then R divides the set S_e into $NG = 2^a$ equivalence classes. Each class is presented by $[^aGe_i]$, where $a \in \mathbb{N}$ and $i \in NG$. In the same manner, R divides the set $S_o = \{1, 3, 5, 7, ...\}$ into NG equivalence classes $[^aGo_i]$.

Example 7. Based on Def. 5.19, R builds the following equivalence classes on the set S_e

$$a = 1 \quad NG = 1 \quad [^1Ge_1] = [0] = \{0, 2, 4, ...\},$$
$$a = 2 \quad NG = 2 \quad [^2Ge_1] = [0] = \{0, 4, 8, ...\}, \quad [^2Ge_2] = [2] = \{2, 6, 10, ..\},$$
$$a = 3 \quad NG = 4 \quad [^3Ge_1] = [0] = \{0, 8, 16, ...\} \quad [^3Ge_2] = [2] = \{2, 10, 18, ..\},$$
$$[^3Ge_3] = [4] = \{4, 12, 20, ...\} \quad [^3Ge_4] = [6] = \{6, 14, 22, ...\}.$$

The next proposition describes the effects of time translations on the outcomes of the BNS algorithm.

Proposition 2. Let the shifted versions of signal $x[n]$ be given by $x[n - S_{e/o}]$, where S_e and S_o are even and odd shifts. The best-detected nodes by Algorithm 4 for the signal $x[n]$ are represented by $^{s}BQ_{\mathbb{C},i}$ and $^{s}BR_{\mathbb{C},i}$, and similarly for $x[n - S_{e/o}]$ by $^{s}BQ'_{\mathbb{C},i}$ and $^{s}BR'_{\mathbb{C},i}$. Considering Def. 5.19, the equivalence relation R builds NG equivalence classes in each scale s of the ACQTWP transform. The impact of the time translation on the best nodes is illustrated by:

- $\forall\ x[n - S_e],$

$$
\begin{cases}
\textbf{a.}\ \forall\ S_e \in [{}^sGe_1], \\
\quad {}^sBQ'_\mathbb{C} = {}^sBQ_\mathbb{C}, \quad {}^sBR'_\mathbb{C} = {}^sBR_\mathbb{C} \\[2mm]
\textbf{b.}\ \forall\ S_e \in [{}^sGe_i]\ 1 < i \le NG, \\
\quad {}^sBQ'_\mathbb{C} = {}^sBQ_{\mathbb{C},(i+(S_e \bmod\ 2^{s-1}))}, \quad {}^sBR'_\mathbb{C} = {}^sBR_{\mathbb{C},(i+(S_e \bmod\ 2^{s-1}))}.
\end{cases}
$$
(5.11)

- $\forall\ x[n - S_o],$

$$
\begin{cases}
\textbf{a.}\ \forall\ S_o \in [{}^sGo_1], \\
\quad {}^sBQ'_{\mathbb{C},i} = {}^sBQ_{\mathbb{C},(i+1)}, \quad {}^sBR'_{\mathbb{C},i} = {}^sBR_{\mathbb{C},(i+1)} \\[2mm]
\textbf{b.}\ \forall\ S_o \in [{}^sGo_i]\ 1 < i \le NG, \\
\quad {}^sBQ'_{\mathbb{C},i} = {}^sBQ_{\mathbb{C},(i+(S_o \bmod\ 2^{s-1}))}, \quad {}^sBR'_{\mathbb{C},i} = {}^sBR_{\mathbb{C},(i+(S_o \bmod\ 2^{s-1}))}.
\end{cases}
$$
(5.12)

Proof. According to the structure of the ACQTWP transform, after low- and high-pass filtering, even and odd downsamplers decompose a shifted and non-shifted version of the input signal in each scale. Def. 5.19 is another way to define the downsamplers $\downarrow 2^e$ and $\downarrow 2^o$. In each scale, the set of even shifts S_e is divided by the downsamplers $\downarrow 2^e$ into 2^s equivalence classes. In the case of even shifts S_e, the nodes in each scale s are categorised into the stated equivalence classes, such as:

$s = 1,\ {}^1Q_{\mathbb{C},0}$ is assigned to $[{}^1Ge_1]$ since $S_e \in [{}^1Ge_1]$.

$s = 2,\ {}^2Q_{\mathbb{C},0},\ {}^2Q_{\mathbb{C},1}$ are respectively assigned to $[{}^2Ge_1], [{}^2Ge_2]$

since $S_e \in [{}^2Ge_1], S_e \in [{}^2Ge_2]$.

$s = 3,\ {}^3Q_{\mathbb{C},0},\ {}^3Q_{\mathbb{C},1},\ {}^3Q_{\mathbb{C},2},\ {}^3Q_{\mathbb{C},3}$ are respectively assigned to $[{}^3Ge_1], [{}^3Ge_2],$

$[{}^3Ge_3], [{}^3Ge_4]$ since $S_e \in [{}^3Ge_1], S_e \in [{}^3Ge_2], S_e \in [{}^3Ge_3], S_e \in [{}^3Ge_4]$.

Depending on S_e and scale s, the best nodes detected by Algorithm 4 for the shifted versions of the signal are not always the same.

The same holds for $\downarrow 2^o$ and odd shifts S_o. Consequently, the best nodes for the shifted versions of signal $x[n]$ are located in nodes with the index $S_{e/o}\ \bmod\ 2^{(s_b - 1)}$.
□

The issues explained in Proposition 2 are illustrated in the following example.

Example 8. The second scale is selected as the best scale for the three motifs in Example 4 ($s_b = 2$) based on the early abandoning algorithm. According to the noble identities [Vai90], the downsampler in the second scale is equivalent to $\downarrow 4^e$. In the second scale of the ACQTWP transform, two even downsamplers select the shifted and non-shifted version of the signal. Employing Def. 5.19 for an even shift

S_e in signal $m_2 = m_1[n - 2548]$ leads to

$$\begin{cases} \text{if } (S_e \quad \text{mod } 4) = 0, & S_e \in [^2Ge_1], \\ \text{otherwise,} & S_e \in [^2Ge_2]. \end{cases}$$

This means that in the second scale, the approximation coefficients of motif m_1 are identical to one of the approximation coefficients nodes of $m_2 = m_1[n - S_e]$ because of the two even downsamplers:

$$\begin{cases} ^2Q'_{\mathbb{C},0}[n] = {}^2Q_{\mathbb{C},0}[n], \\ ^2Q'_{\mathbb{C},1}[n] = {}^2Q_{\mathbb{C},1}[n - \lceil \frac{S_e}{4} \rceil], \end{cases}$$

where $^2Q_{\mathbb{C},0}$ and $^2Q'_{\mathbb{C},0}$ depict the approximation coefficients of m_1 and its shifted version m_2, respectively. Consequently, based on Proposition 2,

$$\begin{cases} ^2BQ'_0 = {}^2BQ_0, \\ ^2BQ'_1 = {}^2BQ_1. \end{cases}$$

Signal $m_5 = m_1[n-1027]$ and the odd shift S_o follow the same procedure. For signal $m_2 = m_1[n - 2548]$, $2548 \in [^2Ge_1]$ and signal $m_5 = m_1[n - 1027]$, $1027 \in [^2Go_1]$, therefore, based on Proposition 2, the best-detected nodes for m_1 and m_2 are similar. On the other hand, the best nodes for signal m_5 are different from m_1.

The best nodes determined by the BNS algorithm vary if the signal is shifted. Nevertheless, the best nodes of the input signal and its shifted versions are all located on the same scale. The successive section introduces the next step in the KITE method: feature extraction.

5.4 Feature Extraction from Variable Scales

Up to this point, the input signal is segmented into various subsequences and analysed by the ACQTWP transform in the representation step, resulting in coefficients of the two selected nodes. Despite the advantages of the ACQTWP transform (cf. Sec. 5.3), this transformation is not invariant to, e.g. reflection. Additionally, the outcomes of this transformation vary for signals with different lengths. In order to identify such signals, KITE employs feature extraction at this stage by obtaining features from the coefficients of nodes selected by the BNS algorithm.

Before extracting features, coefficients of each chosen node are normalised. This is performed since the best-selected scale s_b, and consequently, nodes $^{s_b}BQ_{\mathbb{C},i}$ and $^{s_b}BR_{\mathbb{C},i}$ may vary for each signal. Thus, normalisation provides the ability to compare extracted features from different scales and nodes.

From various normalisation methods [PoM07], the Min-Max method is applied here, as its performance is not related to the type or the distribution of the data. Additionally, it is invariant under uniform scaling mapping (it is proportion-invariant).

Definition 5.20 (Min-Max normalisation [PoM07])**.** Normalised complex approximation coefficients of ACQTWP are given by

$$\widetilde{{}^sQ_{\mathbb{C}}}[n] = \frac{{}^sQ_{\mathbb{C}}[n] - \min({}^sQ_{\mathbb{C}}[n])}{\max({}^sQ_{\mathbb{C}}[n]) - \min({}^sQ_{\mathbb{C}}[n])}. \tag{5.13}$$

Likewise, complex detail coefficients of the ACQTWP transform are normalised by replacing ${}^sR_{\mathbb{C}}[n]$ with ${}^sQ_{\mathbb{C}}[n]$.

Lemma 5. Min-Max normalisation is proportion-invariant so that for $\alpha \in \mathbb{R}$, and complex approximation coefficients ${}^sQ_{\mathbb{C}}$ and $\alpha \cdot {}^s Q_{\mathbb{C}}$, the following holds:

$$\widetilde{{}^sQ_{\mathbb{C}}} \cong \widetilde{\alpha \cdot {}^s Q_{\mathbb{C}}}.$$

The same applies for complex detail coefficients.

Proof. Let ${}^sQ_{\mathbb{C}} = (\mathbb{C}_1, ..., \mathbb{C}_N)^T$ and $\alpha \cdot {}^s Q_{\mathbb{C}} = (\alpha \cdot \mathbb{C}_1, ..., \alpha \cdot \mathbb{C}_N)^T$ so that ${}^sQ_{\mathbb{C}} \cong \alpha\,{}^sQ_{\mathbb{C}}$. It is known that if a signal such as ${}^sQ_{\mathbb{C}}$ is multiplied or divided by a constant like $\alpha \in \mathbb{R}$, the properties of ${}^sQ_{\mathbb{C}}$ such as min, max, mean, median, and the standard deviation are also multiplied or divided by that factor. So,

$$\frac{{}^sQ_{\mathbb{C}} - \min({}^sQ_{\mathbb{C}})}{\max({}^sQ_{\mathbb{C}}) - \min({}^sQ_{\mathbb{C}})} = \frac{\alpha\,{}^sQ_{\mathbb{C}} - \min(\alpha\,{}^sQ_{\mathbb{C}})}{\max(\alpha\,{}^sQ_{\mathbb{C}}) - \min(\alpha\,{}^sQ_{\mathbb{C}})},$$

and $\widetilde{{}^sQ_{\mathbb{C}}} \cong \widetilde{\alpha\,{}^sQ_{\mathbb{C}}}$. $\qquad\square$

After normalising the coefficients of the selected nodes $\widetilde{{}^sBQ_i}$ and $\widetilde{{}^sBR_i}$, where $i \in \mathbb{N}$ is the index of the nodes and $s = s_{\mathrm{b}}$, feature extraction begins. Feature extraction is mainly an application-based task, but in order to generalise KITE's approach, features must be selected independently of the applications under investigation. The chosen set of features must represent the characteristics of the signals, such as shape, variability, and central tendency, to assist motif discovery. Moreover, these features must be invariant to reflection (time-reversal) mapping in order to detect ill-known motifs that are altered by reflection mapping.

Such features are statistical moments, like the arithmetic mean value μ, variance σ, skewness γ, and kurtosis β [PoM07], which have proved their performance in many applications [BDM$^+$12, WGG$^+$15, SYS$^+$15, GMVEFC16, SKR17, BDD$^+$19, ASC19]. This collection of features characterises the shape of a distribution. However, these features alone cannot provide a comprehensive overview of the distribution, but when combined, they are able to determine a quite accurate overview [Blu09]. Besides being invariant to reflection mapping, these four features are translation invariant, leading to support the shift-invariant motif discovery by KITE. These issues are explained in the next Lemmas.

Lemma 6. If $x[n]$ is altered by the reflection mapping such that $x^{\mathrm{R}}[n]$ is the time-reversed version of $x[n]$ ($x^{\mathrm{R}}[n] = x[-n]$), then ${}^sQ[n]$ and ${}^sQ^{\mathrm{R}}[n]$ are the approximation coefficients of signal $x[n]$ and its time-reversed. For extracted feature arrays $f_x[n]$ and $f_{x^R}[n]$ that include the four statistical features $(\mu, \sigma, \gamma, \beta)$, the following holds

$$f_x[n] \cong f_{x^R}[n].$$

Proof. The ACQTWP is not invariant to reflection mapping, meaning that if signal $x^R[n]$ is the time-reversed of signal $x[n]$, then its coefficients, denoted by $^sQ^R[n]$, are also time-reversed version of $^sQ[n]$. According to Def. 5.6, if $x^R[n] = x[-n]$, then for ACQTWP's approximation and detail coefficients obtained from signals $x^R[n]$ and $x[n]$, the following holds

$$^sQ^R[n] \cong {}^sQ[-n] \quad \text{and} \quad {}^sR^R[n] \cong {}^sR[-n].$$

Consequently, the four statistical moments $\mu, \sigma, \gamma, \beta$ obtained from $^sQ^R[n]$ and $^sQ[-n]$ are equivalent. The same holds for $^sR^R[n]$ and $^sR[-n]$ resulting in $f_x[n] \cong f_{x^R}[n]$.   □

Lemma 7. Let $x_1 = x[n]$ and $x_2 = x[n - S_{e/o}]$, where $S_{e/o} \in \mathbb{Z}$, and the two signals x_1, x_2 belong to the equivalence class $[x]$. Feature arrays $f_{x_1}[n]$ and $f_{x_2}[n]$ contain the four statistical features $(\mu, \sigma, \gamma, \beta)$ extracted from coefficients of the selected nodes determined by the BNS algorithm. Then for

$$x_1, x_2 \in [x] \Rightarrow f_{x_1}[n] \cong f_{x_2}[n].$$

Proof. Based on Corollary 2, the ACQTWP transform is shift invariant, and the coefficients of the signal and its shifted ones are equivalent. Thus,

$$\forall x_2, x_1 \in [x] \Rightarrow \begin{cases} {}^sQ'[n] \cong {}^sQ[n - \kappa], \\ {}^sR'[n] \cong {}^sR[n - \kappa], \end{cases} \Rightarrow \begin{cases} {}^sQ'[n], {}^sQ[n] \in [Q], \\ {}^sR'[n], {}^sR[n] \in [R], \end{cases} \quad (5.14)$$

where $^sQ[n]$ and $^sQ'[n]$ are ACQTWP's approximation coefficients of signals x_1 and x_2, which belong to the equivalence class $[Q]$. Similarly, $^sR[n]$ and $^sR'[n]$ are the detail coefficients, which are members of the equivalence class $[R]$. The constant $\kappa \in \mathbb{Z}$ is obtained by $\kappa = \lceil \frac{S_{e/o}}{2^s} \rceil$, where s is the number of scales.
It is well-known that the n-th central moment is shift invariant [Mck19], and so the four statistical features extracted from the ACQTWP's coefficients, as in Eq. (5.14), are shift invariant and equivalent, $f_{x_1}[n] \cong f_{x_2}[n]$.   □

In addition to the four stated features, the maximum and minimum of the phase of the wavelet coefficients are extracted since the phase spectrum provides the characteristic of the original signal and describes its behaviour [ZWS09]. It indicates the location of singularity or irregularity in signals. By utilising the phase information, different kinds of transition points of the analysed signal, such as local maxima and inflection points, can be distinguished [OpL81, Che09]. As a result, selecting this feature makes complex wavelet transformations superior to real wavelets [ToL15].

Definition 5.21 (Phase of Coefficient). Assume $^sBQ_{\mathbb{C},j}[n]$ are the complex approximation coefficients of the best nodes $^sBQ_{\mathbb{C},j}[n] = {}^sBQ_{\mathrm{A},j}[n] + i{}^sBQ_{\mathrm{B},j}[n]$, with the magnitude

$$\mid {}^s BQ_{\mathrm{C},j}\mid = \sqrt{({}^s BQ_{\mathrm{A},j})^2 + ({}^s BQ_{\mathrm{B},j})^2}.$$

Then, when $\mid {}^s BQ_{\mathrm{C},j}\mid > 0$, the phase of ACQTWP's approximation coefficients in each scale is given by ${}^s\theta_Q$ and obtained by

$$ {}^s\theta_Q = \arctan\left(\frac{{}^s BQ_{\mathrm{B},j}[n]}{{}^s BQ_{\mathrm{A},j}[n]}\right), \tag{5.15}$$

where ${}^s BQ_{\mathrm{B},j}[n]$ and ${}^s BQ_{\mathrm{A},j}[n]$ are the approximation coefficients of the best-selected nodes from WPT A and WPT B, in scale s_{b}, and $0 \le j < 4^s$ is the index of nodes. The phase of the detail coefficients is computed similarly.

The maximum and minimum phases are denoted by $\max({}^s\theta_Q)$ and $\min({}^s\theta_Q)$.

It should be noted that extracting the most informative features is challenging for motif discovery due to the unavailability of information about the data and motifs in advance. However, KITE's feature extraction is flexible, meaning that features can be added to or omitted from this step. For instance, for detecting motifs in vibration signals, the most commonly-applied features [BDM+12,REJ+17,FJH+18, YKA+19] for these signals, such as zero-crossing or root mean square [Tho65], can be added to KITE's features. This issue is explained in Chapter 6.

After extracting the aforementioned features from the selected nodes of the ACQTWP transform, KITE determines the similarity between signals, as clarified in the following section.

5.5 Threshold Determination for Similarity Detection

Based on KITE's previous steps, all the segmented subsequences of equal or variable length are transformed into a set of features. In this stage, to differentiate and detect motifs, the similarity between features must be obtained.

According to Def. 1.1, two subsequences are matched and considered as motifs if $|dist(m_1, m_2)| \le \tau$, where threshold τ must be $0 \le \tau \le 1$ (since $0 \le |dist(m_1, m_2)| \le 1$). The full match between two subsequences is obtained when $|dist(m_1, m_2)| = 0$, and the highest dissimilarity is given by 1.

KITE applies threshold τ and a distance measure (cf. Sec. 3.3) to quantify the similarity between the features and determine motifs. This process is depicted in Algorithm 5. In Algorithm 5, the similarity between subsequences is obtained and denoted by the similarity degree $SD[n]$. Before comparing the computed similarity degrees with the threshold τ, $SD[n]$ is normalised by means of Max-Min normalisation (Eq. 5.13).

Normalisation is performed since the outcome of some distance measures, such as Euclidean distance or Edit distance, is not bounded between 0 and 1. After that, similarity degrees are compared to threshold τ in order to detect motifs.

Algorithm 5 Similarity detection
Input: features of all the subsequences f_{s_i}, where $1 \leq i \leq N_s$ (N_s is number of all subsequences); similarity threshold τ.
Output: similarity degrees SD and equivalence class of motif $[m_i]$

1: **for** $i = 1 : 1 : N_s$ **do**
2:   **for** $j = 1 : 1 : N_s$ **do**
3:     $SD(i,j) = |dist(f_{s_i}, f_{s_j})|$
4:   **end for**
5:   $\widetilde{SD}(i,:) = \text{normalise}(SD(i,:))$
6:   $[m_i] = find(SD(i,:) <= \tau)$
7: **end for**

Determining an optimal threshold for motif discovery is challenging, as motif discovery is an unsupervised task [ToL18]. In this work, threshold τ is either provided by the user, as in [BuT01, PKL$^+$02, FAS$^+$06, CaA10, ASS$^+$15, BWP16], or defined automatically.

Choosing a large threshold results in accepting improper matches, and a small threshold discards possible motifs. Thus, threshold τ must be defined so that it minimizes the amount of false-positive and false-negative motifs. The determination of τ is based on the frequency distribution of the quantified similarities. A frequency distribution, illustrated by a histogram, summarises data by dividing it into mutually exclusive groups and the number of occurrences in a group.

Definition 5.22 (Histogram of the similarity degrees). The histogram of similarity degrees is denoted by H_{SD} and given by $H_{SD} = \sum_{i=1}^{b} h(i)$, where $b \in \mathbb{N}$ is the number of histogram bins, and $h(i)$ is the proportion in each bin. Each bin has a lower and upper edge denoted by LE and UE (LE, $UE \in \mathbb{R}$), respectively.

Next, a three-step approach is proposed to determine the number of bins for histogram H_{SD}:

1. Given the similarity degree $\widetilde{SD}[n]$, the absolute value of the differences between all successive similarity degrees in $\widetilde{SD}[n]$ is computed by

$$\widetilde{SD}_{\text{dif}}[n] = |\widetilde{SD}[i] - \widetilde{SD}[i+1]|, \ \forall \ 1 \leq i \leq N-1,$$

   where N is the length of $\widetilde{SD}[n]$. This shows the separation and departure of the values in $\widetilde{SD}[n]$.

2. The standard deviation $\sigma(\widetilde{SD}_{\text{dif}})$ is considered as the width for the histogram bins since it indicates how outspread the data are. A small amount of standard deviation $\sigma(\widetilde{SD}_{\text{dif}})$ depicts a higher concentration in data samples, and therefore, a larger number of bins with a smaller width is required. On the contrary, a high value of $\sigma(\widetilde{SD}_{\text{dif}})$ indicates a lower concentration for data samples, resulting in fewer bins with a broader width.

3. The number of bins is attained by dividing the interval $[0, 1]$ by the standard
   deviation $\sigma(\widetilde{SD}_{\text{dif}})$, $b = \frac{1}{\sigma(\widetilde{SD}_{\text{dif}})}$.

After determining the bins for histogram H_{SD}, threshold τ is obtained based on
the global (or local) maximum of the histogram H_{SD}. Each bin of the histogram
H_{SD} represents a group of subsequences whose similarities are in the same range.
The global maximum of the histogram H_{SD} is the bin with the highest density
(number of similar subsequences). The global maximum is only considered if it
belongs to bins in the interval of $[0, 0.5]$ since threshold τ must be $0 \leq \tau \leq 0.5$.
Assigning $\tau > 0.5$ allows subsequences as motifs with a higher dissimilarity, thereby
increasing the false-positive rate.
If the global maximum is not in the interval of $[0, 0.5]$, then the local maximum
in this interval must be selected. Next, the upper bound of the global (or local)
maximum of the histogram H_{SD} is defined as the threshold τ and is given by the
following:

Definition 5.23 (Threshold). Let $\mathbb{B} = \{1, 2, ..., n\}$, for $n < b$ be a set of bins whose
upper edges are less than 0.5, $UE_1 < UE_2 < ... < UE_n \leq 0.5$. Then, threshold τ
is computed by

$$\tau = \begin{cases} UE_i & \text{for } \max\limits_{i=1}^{n} h(i), \\ 0.5 & \text{otherwise.} \end{cases} \tag{5.16}$$

Thus, threshold τ is equal to the upper edge of the bin $i \in \mathbb{B}$ with the maximum
proportion. If all the bins in $\mathbb{B}$ are uniformly distributed, then $\tau = 0.5$.
Note: If the number of local maximums of H_{SD} in the interval $[0, 0.5]$ is more
than one, then the largest upper bound must be set to threshold τ.
The following example explains the aforementioned procedure.

Example 9.   Consider signal $x[n]$ and its 12 segmented subsequences by sliding
window $w[n]$, as demonstrated in Fig. 5.9.

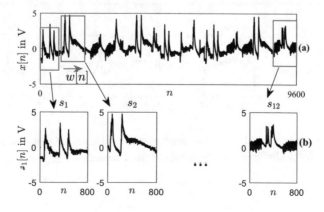

Figure 5.9: (a) Signal $x[n]$ and the sliding window $w[n]$, (b) Segmented subsequences from signal $x[n]$.

After KITE's representation and feature extraction steps, the similarity degrees for all of the subsequences are obtained. For subsequence s_1, these similarity measurements are given in $\widetilde{SD} = [0, 0.37, 0.71, 1, 0.38, 0.60, 0.34, 0.45, 0.18, 0.59, 0.44, 0.78]$. Based on the explained approach, $\sigma(\widetilde{SD}_{\text{dif}}) = 0.137$ and so $b = 7$. Histogram H_{SD} is depicted in Fig. 5.10.

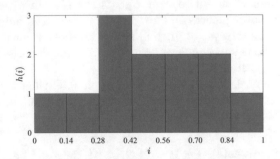

Figure 5.10: Histogram of H_{SD} for subsequence $s_1[n]$, the width of each bin is set to 0.137, and the number of bins is equal to 7.

Based on Eq. 5.16, the threshold is set to $\tau = 0.42$, and subsequences with similarity degrees smaller than $\tau = 0.42$ are considered as motifs for subsequence s_1, presented in Fig. 5.11.

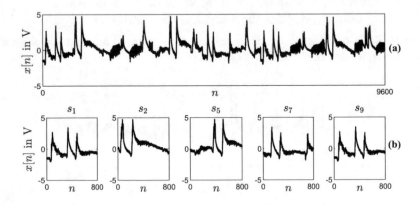

Figure 5.11: Detected motifs for subsequence s_1. Subsequences s_2, s_5, s_7, and s_9 have a similarity degree less than threshold $\tau = 0.42$.

5.6 Significant Motif Discovery

After detecting similarities between subsequences, several candidate motifs are obtained. Commonly, it is the expert's task to evaluate and select the most beneficial motifs since not all the obtained motifs contain valuable and meaningful information (cf. Example 10). For this reason, significant motif discovery is proposed.

Definition 5.24 (Significant motifs). Significant motifs are selected motifs that assist the domain expert in discriminating unnecessary or random similarities. Such motifs are considered as a match without having huge overlapping. Additionally, they provide a pair of most similar motifs in each detected motif equivalence class.

Example 10. Some discovered motifs by KITE, from the signal in Example 2 (depicted in Fig. 5.12 (a)), are illustrated in Fig. 5.12 (b)-(e).

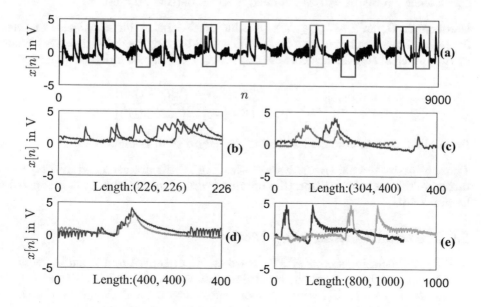

Figure 5.12: (a) Signal $x[n]$ is segmented into subsequences of variable lengths. Examples of detected motifs: (b,d) motif pairs with equal lengths; (c,e) motif pairs with various lengths.

For transparency, not all the detected motifs are given in Fig. 5.12. Motif pairs that are depicted in sub-figures (b) and (d) have equal lengths. On the contrary, motif pairs in sub-figures (c) and (e) have variable lengths. Among all these motifs, some may be unattractive motifs, and some may result in appealing information for entomologists, e.g., the motif pair depicted in sub-figure (d) shows a specific injection process [Mue14].

Similar to other motif discovery methods, KITE also results in a pool of various motifs, especially if motif lengths are not determined, as explained in the two following Lemmas.

Lemma 8. By providing signal $x[n]$ of length $N \in \mathbb{N}$, the motif's length l_d, and the overlapping degree O_d, KITE results in R number of equivalence classes $[m_i]$, where $i \in \mathbb{N}$ is the index of motif m_i and

$$0 \leq R \leq \cdot \left\lceil \frac{N - (l_d - 1)}{2 \cdot (\lfloor l_d - (l_d \cdot O_d) \rfloor)} \right\rceil.$$

Proof. The number of subsequences obtained after employing Algorithm 1 is equal to $\left\lceil \frac{N - (l_d - 1)}{(\lfloor l_d - (l_d \cdot O_d) \rfloor)} \right\rceil$. From all these subsequences, the maximum number of detected equivalence classes $[m_i]$, which is given by $\left\lceil \frac{1}{2} \cdot \frac{N - (l_d - 1)}{(\lfloor l_d - (l_d \cdot O_d) \rfloor)} \right\rceil$ is achieved if each subsequence has only one match $((Card([m_i])) = 2)$. The minimum number of equivalence classes, $R = 0$, is obtained if KITE detects no motif. □

Lemma 9. Given $2 \leq l_{nd} \leq N/2$ possible motif lengths for signal $x[n]$ of length $N \in \mathbb{N}$ and overlapping degree O_d, KITE's variable lengths motif discovery results in R number of equivalence classes $[m_i]$, where $i \in \mathbb{N}$ is the index of motif m_i and

$$0 \leq R \leq \sum_{i=1}^{Card(l_{nd})} \left\lceil \frac{1}{2} \cdot \frac{N - (l_{nd_i} - 1)}{\lfloor l_{nd_i} - (l_{nd_i} \cdot O_d) \rfloor} \right\rceil.$$

Proof. Lemma 9 is proved in a similar procedure as in Lemma 8. □

Thus, to assist the expert analysis, misleading motifs are excluded and representative motifs are selected from the obtained equivalence classes of motifs given in Lemmas 8-9.

5.6.1 Excluding Misleading Motifs

In the pre-processing step of KITE, based on Algorithms 1 and 2, sliding windows of different lengths segment the input data into subsequences with diverse lengths. Given O_d, these algorithms obtain more subsequences, and therefore, the chance of discovering more motifs is increased. As a result, successive segments or subsequences with a vast overlap have the potential to be detected as motifs. Typically, such motifs occur if the length of the sliding window is increased by, e.g., one. In this case, two subsequences share exactly $w - 1$ points, where $w \in \mathbb{N}$ is the length of the sliding window. Therefore, these two subsequences are held to be very similar since they have considerable overlap. These types of motifs are called *trivial motifs* [PKL+02, GSST16].
Note: It should be noted that the term trivial motifs is commonly applied in the motif discovery community. Nevertheless, the author decided to use the term *misleading motifs* instead of trivial motifs since the word trivial can be defined broadly.

Definition 5.25 (Misleading motifs). Let $s_{i,w} = (s_i, ..., s_{i+w-1})$ and $s_{j,w} = (s_j, ..., s_{j+w-1})$ of length $w \in \mathbb{N}$ with the first position at i and $j \in \mathbb{N}$ be two matched subsequences or motifs. Subsequence $s_{j,w}$ is a trivial match to subsequence $s_{i,w}$ if $i = j$ or $\frac{|s_{i,w} \cap s_{j,w}|}{w} > O_d$.

The first step to select meaningful motifs is to exclude misleading motifs by providing an offset and incrementing the sliding window by that offset to avoid such misguiding matched motifs, as in Algorithm 1. The overlapping degree O_d, Def. 5.3, is an example of such an offset. KITE's approach towards excluding misleading motifs is divided into equal- and variable-length motif discovery.

1. Misleading motifs exclusion for equal-length motif discovery:

Lemma 10. For non-overlapping, equal-length motif discovery, if the motif's length l_d is given, then no misleading motifs are among KITE's detected motifs.

Proof. Without any overlapping, Algorithm 1 segments signals by sliding windows with offset l_d. □

Lemma 11. If the length of the motif l_d and the overlapping degree O_d are given, then none of the detected motifs of equal length by KITE is a misleading motif pair.

Proof. Based on Remark 1, the total number of $\lceil \frac{N-(l_d-1)}{[l_d-(l_d \cdot O_d)]} \rceil$ subsequences from the signal $x[n]$ is obtained by employing Algorithm 1. Assume all these subsequences are motif pairs, then two subsequences $s_1 = (x_i, ..., x_p, ..., x_{i+j-1})$ and $s_2 = (x_p, ..., x_j, ..., x_{p+q-1})$ of lengths $j, q \in \mathbb{N}$ belong to the equivalence class $[m_1]$, if $|dist(f_1, f_2)| \leq \tau$, then $s_1, s_2 \in [m_1]$, where f_1 and f_2 (cf. Sec. 5.4) are extracted features from s_1 and s_2. Then regardless of their overlapping degree, these two subsequences are not a misleading motif pair, since the overlapping degree between them cannot be larger than the given O_d, thus $\frac{|s_1 \cap s_2|}{max(j,q)} \not> O_d$. □

2. Misleading motifs exclusion for variable-length motif discovery:
For variable-length motif discovery, Algorithm 2 divides the signal $x[n]$ of length $N \in \mathbb{N}$ into various subsequences of different lengths (e.g. from 2 samples up to $N/2$). Regardless of the overlapping degree O_d, some of the subsequences obtained from Algorithm 2 have more common segments.
A set of all employed sliding windows in Algorithm 2 is denoted by W and defined as

$$W = \{w_1[n], w_2[n], ..., w_j[n]\}, \text{ where } j = ((N-1)/2) \text{ and } 2 \leq \overset{j}{\underset{i=1}{length}}(w_i) \leq N/2,$$

so that $w_1[n] \subset w_2[n] \subset ... \subset w_j[n]$. Consequently, subsequences obtained by each of these sliding windows have common sections.
For identifying non-misleading motifs of variable lengths, the J_d distance is proposed, which computes the diversity between two motifs. The J_d distance employs the concept of Jaccard similarity or the so-called Jaccard index (original

paper [Jac01]), which is mostly applied to compare the similarity and variety of two sets.

Proposition 3 (J_d distance). Assume a pair of motifs $m_1, m_2 \in [m_1]$, so that $m_1 = (m_i, ..., m_p)$ and $m_2 = (m_j, ..., m_q)$ of length $l_1, l_2 \in \mathbb{N}$, where i, j are the first position, and p, q are the last position of the motifs. Then [Jac01]

$$J_d(m_1, m_2) = \frac{(l_1 + l_2) - (\max(p, q) - \min(i, j)) - 1}{\max(l_1, l_2)}, \qquad (5.17)$$

where $J_d(m_1, m_2) \leq 1$. If $J_d \leq 0$, then two motifs have no common segments. On the other hand, two motifs share some segments if $J_d > 0$, and in the case of $J_d = 1$, two subsequences are entirely overlapped [ToL17a].

By applying J_d distance, the non-misleading variable-length motifs are determined, as given in Algorithm 6. If overlapping is accepted, then non-misleading motifs are detected when $J_d \leq O_d$. Alternatively, if overlapping is not required, then motifs with $J_d \leq 0$ are considered non-misleading.

Algorithm 6 Excluding misleading motifs of variable length
Input: motifs $m_{i,p}, m_{j,q}$, overlapping degree: O_d
Output: non-misleading motifs

1: $Min\_Pos = min(i, j)$;
2: $Max\_Pos = max(p, q)$;
3: $l_1 = length(m_{i,p})$;
4: $l_2 = length(m_{l,q})$;
5: $J_d = \frac{sum(l_1, l_2) - (Max\_Pos - Min\_Pos)}{Max\_Pos}$;
6: **if** $J_d <= O_d$ **then**
7:     $m_{i,p}$ and $m_{j,q}$ are not a misleading motif pair.;
8: **else**
9:     return no motifs.;
10: **end if**

5.6.2 Representative Motifs

Besides excluding misleading motifs, detecting *representative* motifs also supports the domain expert in selecting significant motifs.

The definition of the nearest-neighbour and K-frequent motifs is given in Chapter 2, Def. 2.8 and 2.10. These definitions are practical for motifs of equal lengths and not for ill-known motifs, and therefore, they are improper for KITE. However, by combining these definitions, it is possible to deliver representative motifs.

Definition 5.26 (*K-representative motif*). The K-representative motif is denoted by $^K m_R$ and includes the two motifs with maximum similarity (with minimum distance) in the equivalence class of K-frequent motif $[^K m]$, so

$$\forall m_i, m_j \in [^K m],\ 1 \le i, j \le N,\ i \ne j,\ {}^K m_\mathrm{R} = \min_{i,j}(dist(m_i, m_j)).$$

Thus, 1-representative motif pair, $^1 m_\mathrm{R}$, has the largest similarity in the equivalence class of 1-frequent motif $[^1 m]$,

$$\forall m_i, m_j \in [^1 m],\ 1 \le i, j \le N,\ i \ne j,\ {}^1 m_\mathrm{R} = \min_{i,j}(dist(m_i, m_j)),$$

where $N \in \mathbb{N}$ is the number of motifs in $[^1 m]$.

It must be noted that $\mathrm{Card}([^1 m]) = \max_K(\mathrm{Card}([m_K]))$ and $K \in \mathbb{N}$ is the number of all discovered motif classes. Thus, the equivalence class $[^1 m]$ has the highest amount of detected motifs.

Algorithm 7 delivers the K-representative motifs of signal $x[n]$.

Algorithm 7 Representative Motifs

Input: equivalence classes of K-frequent motifs $[^1 m], [^2 m], ..., [^K m]$, similarity degrees of K-frequent motif classes $^1 SD, {}^2 SD, ..., {}^K SD$

Output: K-representative motifs: $^1 m_\mathrm{R}, ..., {}^K m_\mathrm{R}$

1: **for** $i = 1 : K$ **do**
2:      $M_i = [^i m]$;
3:      $(-, Pos) = \min(^i SD)$
4:      $^i m_R = (M_i(1), M_i(Pos))$;
5: **end for**

KITE provides significant motifs by the two proposed Algorithms 6 and 7 to facilitate selecting the most proper motifs.

Nevertheless, since KITE detects unknown patterns without foreknowledge about them, an expert should still check the (now) known patterns for further usage in tasks such as classification and rule discovery.

5.7 Time Complexity Analysis

Previous sections focus on explaining KITE's structure. In this section, the computational complexity of KITE and its five main steps, pre-processing, representation, feature extraction, similarity measurement, and significant motif detection, are analysed.

1. **Pre-processing:** In this step, time series are segmented into equal- and variable-length subsequences, applying Algorithms 1 and 2.

   Algorithm 1 executes in $\mathcal{O}(N/c) \approx \mathcal{O}(N)$, where the constant c is computed by $c = \frac{1}{(\lfloor l_\mathrm{d} - (l_\mathrm{d} \cdot O_d) \rfloor)}$. The length of the time series $x[n]$ is given by N, l_d and O_d are the pre-defined length for motifs, and the overlapping degree. Algorithm 2 executes in $\mathcal{O}(\frac{N^2}{c}) \approx \mathcal{O}(N^2)$, where $c = \frac{1}{(\lfloor l_\mathrm{nd} - (l_\mathrm{nd} \cdot O_d) \rfloor)}$, N is the length of time series $x[n]$. The overlapping degree is denoted by O_d, $2 \leq l_{nd} \leq N/2$.

2. **Representation:** The ACQTWP transform transforms all of the segmented subsequences to analyse and obtain necessary information. The computational complexity of ACQTWP is given in Lemma 12.

Lemma 12. The filter bank implementation of the ACQTWP transform using equations in Def. 5.5 requires $\mathcal{O}(cN \log_2 N)$ operations, where $c = 4$.

Proof. Consider signal $x[n]$ of length $N \in \mathbb{N}$ and filters ${}^s h_a[n]$ and ${}^s g_a[n]$ of length $M \in \mathbb{N}$. For the wavelet packet WPT A, in the first scale, $2NM$ multiplications and additions are needed to decompose the signal. In the next scale, the signal's length is reduced to $\frac{N}{2}$. Thus, $4\frac{NM}{2}$ multiplications and additions are required. Consequently, the computational complexity of ACQTWP for WPT A is obtained by $(2NM + \frac{4NM}{2} + \frac{8NM}{4} + ...) = 2NM \log_2 N$. As, $M \ll N$, this computational complexity is approximated by $\mathcal{O}(2N \log_2 N)$. The time complexity of the ACQTWP transform is $\mathcal{O}(4N \log_2 N)$, considering both wavelet packet trees. □

As explained in Sec. 5.3.1.3, there is no need to decompose a signal by ACQTWP into the total amount of scales, since the decomposition is aborted by the early abandoning approach (Algorithm 3). Applying this method decreases the time complexity of the ACQTWP transform, as declared in the following Lemma.

Lemma 13. Algorithm 3 stops the wavelet decomposition when the shape of the wavelet coefficients is not similar to the original signal and selects the scale with the maximum amount of information. It reduces the time complexity of the ACQTWP transform to $\mathcal{O}(cN) \approx \mathcal{O}(N)$, where N is the length of time series, and $c = 4s_b$.

Proof. As in Lemma 12, the time complexity is approximately $\mathcal{O}(4N) \approx \mathcal{O}(N)$ in each scale. In Algorithm 3, input signals are analysed up to the selected scale s_b. Computation and comparison of each scale's information content (lines 2-5) are performed in $\mathcal{O}(1)$. Consequently, using Algorithm 3 and for $c = 4s_b$, the time complexity of the ACQTWP transform up to scale s_b is reduced to $\mathcal{O}(cN)$. □

Lemma 14. The BNS algorithm executes in $\mathcal{O}(2^{s_b} N)$.

Proof. BNS detects two nodes with the highest information content from the selected scale s_b in both wavelet packet trees by measuring the energy-to-entropy ratio [ToL18]. The number of nodes in scale s_b is equal to 4^{s_b}, so analysing all these nodes takes $\mathcal{O}(4^{s_b})$. Computing energy-to-entropy ratio (cf. Eq. 5.9) for each node results in $\mathcal{O}(\frac{N}{2^{s_b-1}})$, and this leads to the total time complexity of $\mathcal{O}(2^{s_b} N)$ for BNS. □

3. **Feature extraction:** In the third step of KITE, determining all the features from the two chosen nodes takes $\mathcal{O}(2)$. The four first statistical moments perform in $\mathcal{O}(N/2^{s_b})$, and phase's (Eq. 5.15) time complexity is $\mathcal{O}(\log(N/2^{s_b}))$,

where N is the length of time series $x[n]$, and s_b is the output of Algorithm 3. Consequently, the feature extraction step executes in approximately $\mathcal{O}(N/2^{s_\mathrm{b}})$.

4. **Similarity measure:** The computational complexity of detecting motifs by measuring the similarity between their features is explained below.

Lemma 15. The similarity measure step of KITE approximately has a time complexity of $\mathcal{O}(cN_\mathrm{s}^2)$, N_s is the number of subsequences obtained in the pre-processing step (given in Remarks 1 and 2), and $c = 12$ is the number of features extracted from each subsequence.

Proof. KITE measures the similarities between all equal- and variable-length subsequences by means of Algorithm 5. There are $\mathcal{O}(N_\mathrm{s}^2)$ comparisons between subsequences. Depending on the applied distance measure (e.g., the Euclidean distance performs in $\mathcal{O}(N)$, but the DTW executes in $\mathcal{O}(N^2)$), the minimum computational complexity for each of these comparisons is obtained by $\mathcal{O}(c)$, and the maximum time complexity is given by $\mathcal{O}(c^2)$, where constant $c = 12$ is the number of extracted features for each subsequence. Accordingly, the computational complexity for this step of KITE is approximately attained by $\mathcal{O}(cN_\mathrm{s}^2)$.   □

Remark 3. The time complexity of this step improves by gaining benefit from the symmetry property of a metric distance measure such as Euclidean distance [DeD09, WFBS14]. In that case, the time complexity of $\mathcal{O}\left(\frac{c(N_\mathrm{s}(N_\mathrm{s}-1))}{2}\right)$ is obtained by computing the upper triangular similarity degree matrix instead of the whole matrix.

5. **Significant Motifs:** All detected motifs from the previous step are evaluated by excluding misleading motifs and determining representative motifs.

Lemma 16. Algorithm 6 excludes misleading motifs in $\mathcal{O}(N_\mathrm{m})$, where $N_\mathrm{m} \in \mathbb{N}$ is the number of detected motifs.

Proof. Algorithm 6 eliminates the misleading motifs by computing the diversity between two candidate motifs. All computational functions such as max, min, sum, and division execute linearly. Thus, this algorithm performs in $\mathcal{O}(N_\mathrm{m})$.   □

Lemma 17. Algorithm 7 operates in $\mathcal{O}(K)$, where $K \in \mathbb{N}$ is the number of frequent motifs.

Proof. For all K-frequent motifs, Algorithm 7 performs in $\mathcal{O}(K)$. Operators such as the min operator execute linearly.   □

An overview of the time complexity for each step of KITE is given in Table 5.3. Pre-processing and similarity measurements have the highest computational complexity compared with other steps. In pre-processing, the length of the input data has a significant impact on computational complexity. For the similarity measurement step, the number of subsequences affects the time complexity, rather than the applied distance measures since similarity measurements are employed on features.

Table 5.3: Computational complexity of KITE's five steps. The length of the input signal, number of segmented subsequences, and the total detected motifs are given by $N, N_s, N_m \in \mathbb{N}$. Constants $s_b, c, K \in \mathbb{N}$ are the most informative scale, number of features, and K-frequent motifs.

	Step 1	step 2	Step 3	Step 4	Step 5
Equal	$\mathcal{O}(N)$	$\mathcal{O}(N) + \mathcal{O}(2^{s_b}N)$	$\mathcal{O}(\frac{N}{2^{s_b}})$	$\mathcal{O}(cN_s^2)$	$\mathcal{O}(K) + \mathcal{O}(N_m)$
Variable	$\mathcal{O}(N^2)$	$\mathcal{O}(N) + \mathcal{O}(2^{s_b}N)$	$\mathcal{O}(\frac{N}{2^{s_b}})$	$\mathcal{O}(cN_s^2)$	$\mathcal{O}(K) + \mathcal{O}(N_m)$

As $(s_b, c, K, N_m) \ll N$, and N_s, consequently, KITE's equal-length motif discovery performs in approximately $\mathcal{O}(N + N_s^2)$, and for variable-length motif discovery, KITE's computational complexity is mainly given by $\mathcal{O}(N^2 + N_s^2)$, where N and $N_s \in \mathbb{N}$ are the length of the input signal and the number of segmented subsequences, respectively.

5.8 Summary

KITE is proposed to detect ill-known motifs transformed by translation, uniform scaling, reflection, squeeze and stretch mappings, discover overlaid motifs by noise, and determine motifs of variable lengths. KITE combines both pattern recognition and motif discovery procedures and contains five steps: pre-processing, representation, feature extraction, similarity measurement, and significant motif detection. The signals are divided into subsequences of equal and variable lengths in the pre-processing step. This length plays a vital role in motif discovery and is either provided beforehand or obtained automatically.

In the next step, segmented subsequences are analysed by the ACQTWP transform. This wavelet transformation decomposes signals into a comprehensive time-frequency resolution, reduces the amount of superimposed noise, is approximately analytic and shift-invariant. The distribution and lengths of subsequences have no impact on the performance of the ACQTWP transform. Thus, these properties make ACQTWP transform an appropriate representation tool. As decomposing signals into all ACQTWP transform scales leads to redundant information, the early abandoning algorithm selects a scale based on the information content and shape of the input signal. Subsequently, the best two nodes from the selected scale are considered for feature extraction. These nodes are chosen according to their energy-to-entropy ratio. Regardless of the ACQTWP transform's advantages (cf. Sec. 5.3), this transformation is not invariant to reflection, and its outcomes vary for signals with different lengths. These challenges are handled in the feature extraction step.

This step begins by normalising the coefficients of the two chosen nodes. Normalisation is performed to compare features from various nodes and scales. The maximum and minimum amount of phase and the first four statistical moments are extracted as features since the phase provides the signal's behaviour. Additionally, the four first moments are extracted since they represent the shape characteristics of the signals and are robust against reflection and translation mappings.

The extracted features are assigned to a distance measure to detect motifs by measuring the similarity between the subsequences' features and compare it with a threshold. If the similarity between two subsequences is less than the provided threshold, then the two subsequences are considered a motif. Unlike most motif discovery approaches stated in Chapter 4, this threshold is defined automatically in KITE.

Finally, misleading motifs are excluded, and K-representative motifs are determined. These motifs can later be applied to other data mining tasks, such as classification. The following section provides an evaluation of the proposed method.

6 Evaluation

In the previous sections, the theoretical methods and their properties that contributed to this dissertation have been described. In this chapter, experiments are performed in order to evaluate the performance of KITE. All the examinations are benchmarked against six state-of-the-art algorithms, as explained in Chapter 4. The validation principles that are employed to inspect and compare the performance of KITE are explained in Sec. 6.1. The design of the two types of experiments and investigations is defined in Sec. 6.2. The first type of experiment, explained in Sec. 6.3, aims to detect motifs of equal length, while the second one, given in Sec. 6.4, analyses motifs of variable lengths. For these experiments, it is shown that KITE's outcomes are superior to or in the range of the six state-of-the-art algorithms. Both types of experiments are conducted on several test cases, including synthetic and real-world data.

Additional evaluations on the KITE's robustness towards noise are described in Sec. 6.5, where the test cases are covered with Gaussian noise at signal-to-rate ratio levels of 30dB, 20dB, and 10dB. All of the tested algorithms yield lower results in the existence of noise in the data. Nevertheless, KITE and VALMOD [LZP+18] identified a higher number of correctly detected motifs than other tested methods. The scalability of KITE is examined in Sec. 6.6. It is confirmed that increasing the size of the input data leads to a longer execution time for KITE and other investigated algorithms. Finally, this section concludes by providing a case study of KITE in anomaly detection. The work of the author published in [TDDL16, ToL17a, ToL18] is integrated literally into this section.

6.1 Validation Principles

The best way to compare the results of various methods is to apply the same measure for interpreting the outcomes. This section describes the applied measures that compare KITE's results with the outcomes of the state-of-the-art methods.

6.1.1 Feature Selection

Feature extraction and selection are application-based methods and depend on the data under investigation. It is important not to discard useful information or to include too much information. Thus, in order to choose informative features, methods such as linear discriminant analysis (LDA) [Fis36] (as cited in [Ize08]) are usually employed. As described in Sec. 1.3, feature extraction and selection are not the main focus of this work. Nevertheless, to measure the efficiency of the obtained

© The Author(s), under exclusive license to
Springer-Verlag GmbH, DE, part of Springer Nature 2022
S. Deppe, *Discovery of Ill-Known Motifs in Time Series Data*, Technologien
für die intelligente Automation 15, https://doi.org/10.1007/978-3-662-64215-3_6

features, the LDA method is utilised in this work. LDA quantifies the separability
rate of various motifs according to the extracted features. It projects a data set
into lower dimensional space to maximise class separability [Alp10]. It searches
for a projection where the subsequences belonging to the same motif class can be
linearly classified from the instances of other motif classes [Fis36, Alp10, ToL18].
In order to estimate the merit of the extracted features, the classification error by
LDA is considered. "This error is denoted by e, where $0 \leq e \leq 1$. If the motifs can
be separated linearly and correctly, the error is equal to zero, and if they cannot
be classified linearly and correctly, then the error has its maximum amount of
one" [ToL18].

6.1.2 Quality Measures

Different quality measures are engaged in measuring the performance of KITE.
One of the standard methods for evaluating the result is employing a confusion
matrix [Alp10], which is depicted in Table 6.1. The confusion matrix is based on
four possibilities to qualify a motif m matching a subsequent s_l.

Table 6.1: Confusion Table [Alp10]

True Motif's Class	Predicted Motif's Class		
	Positive	Negative	Total
Positive	tp	fn	p
Negative	fp	tn	n
Total	$\acute{p}$	$\acute{n}$	N

A positive example that is also predicted positive is denoted by tp. The false-
positive rate, fp, represents a positive example that is predicted incorrectly. A
negative example with the prediction which is also negative is represented by tn.
The rate fn describes positive predictions for negative examples [Alp10, ToL18].
The total amount of positive and negative predictions are given by $\acute{p}$ and $\acute{n}$, respectively. In order to benchmark KITE's performance and the tested state-of-the-art
algorithms, using the confusion matrix, the following quality measures are defined:
Correct motif discovery rate CR, Sensitivity Sn, Precision Pr, and F-Measure
$F - M$ [Alp10, WFH11, Pow11].

Definition 6.1 (Correct motif discovery rate). The performance of a motif discovery algorithm is indicated by this measure, which is based on the number of all
motifs ($N \in \mathbb{N}$) and the correctly detected motifs (n^+). This measure is given by
$CR = \frac{n^+}{N}$, where $0 \leq CR \leq 1$ [ToL18].

Definition 6.2 (Sensitivity). Sensitivity, also known as recall, refers to a portion
of subsequences in the target class that are correctly matched by the motif. It
is given by $Sn = \frac{tp}{tp+fn}$, where $0 \leq Sn \leq 1$ and the optimal case is $Sn = 1$
[Pow11, WFH11, ToL15].

Definition 6.3 (Precision). Precision indicates a measure of the target class's subsequences that are matched by the motif and subsequences that are correctly not assigned to the motif. it is obtained by $Pr = \frac{tp}{tp+fp}$, where $0 \leq Pr \leq 1$ [Pow11, WFH11, ToL15]. Thus, "precision relates the number of correctly detected motifs to all positively determined motifs with the optimal case of $Pr = 1$" [ToL18].

Definition 6.4 (F-Measure). F-Measure quantifies the accuracy of the tested method and is a combination of precision and sensitivity. This measure is determined by $F - M = 2 \cdot (\frac{Pr \cdot Sn}{Pr+Sn})$ [Pow11, WFH11].

6.2 Design of the Experiments

As stated in Def. 1.3, the term ill-known refers to motifs that are transformed by affine mappings such as translation and uniform scaling or are altered by stretch and squeeze mappings. Additionally, motifs imposed by noise or which have different lengths are also considered ill-known motifs. In order to evaluate ill-known motif discovery by KITE, two types of experiments are designed. Both experiments follow the same procedure, whereby the first experiment aims to detect motifs of equal length, and the second one focuses on the discovery of variable-length motifs. These examinations are executed following the KITE's structure. In pre-processing, the input data are divided into subsequences of equal and variable lengths by applying the algorithms 1 and 2.

The segmented subsequences are sent to the ACQTWP transformation in the representation step to obtain a comprehensive time-frequency decomposition, illustrated through several scales. Due to information redundancy, the most appropriate scale and nodes are selected based on the proposed methods in Sec. 5.3.1.3. Within this step, ill-known motifs that are covered with noise and are altered by translation, stretch and squeeze mappings are determined. To reduce the size of the data and overcome other problems (e.g. uniform scaling, length variations) that cannot be handled in the representation step, feature extraction is performed. First, features are extracted from the selected nodes, and then the LDA algorithm (cf. Sec. 6.1.1) measures the merit of these features.

After the feature extraction, a distance measure quantifies the similarity between subsequences. The performance of six different distance measures is investigated in this work. Finally, the representative motifs are provided, and misleading ones are rejected. The results of the KITE motif discovery algorithm are benchmarked with six state-of-the-art algorithms, namely the Brute-Force [PKL+02], SCRIMP++ [ZYZ+18], Mr.Motif [CaA10], MOEN algorithm [MuK10], MOEN Enum. [MuC15], and VALMOD [LZP+18]. The Brute-Force [PKL+02] is the first proposed method for the time series motif discovery. SCRIMP++ [ZYZ+18] handles very large data sets ($N > 4000000$). The fastest methods are SCRIMP++ [ZYZ+18] and Mr.Motif [CaA10], a prominent member of the SAX-based methods. The MOEN algorithm [MuK10] is the latest improved version of MK and identifies motifs in on-line applications. MOEN Enum. [MuC15] and VALMOD [LZP+18] belong to the family of methods that can detect motifs of variable lengths. Additionally, VALMOD [LZP$^+$18] provides the top ten detected motifs. VALMOD [LZP$^+$18] is the latest version of motif discovery approaches that apply the matrix profile concept (explained in Chapter 4). All the tests are executed on the test cases as explained in the following section.

6.2.1 Test Cases

Different synthetic and real-world data sets are applied to examine the performance of KITE. The synthetic data are gathered from widely utilised test cases in the domain of time series data mining [CKH$^+$16]. This verifies the evaluation and comparison of the results of KITE with the state-of-the-art approaches. The real-world data are obtained from a research project on monitoring security-critical systems. It should be noted that the evaluation of the performance of KITE and other algorithms with the stated quality measures (Sec. 6.1.2) is possible since the test cases are labelled with the known motif types.

6.2.1.1 Synthetic Data

The data sets are gathered from the UCR time series classification and clustering repository [CKH$^+$16]. They are the most cited and the most commonly-employed data in the domain of time series clustering, classification, and motif discovery. The properties of these data sets are described in Table 6.2.

Table 6.2: Synthetic data sets and their properties.

Data Set	Number of Motif Types	Length of Motifs
Gun Point	1	150
50 Words	25	270
Food	2	286
Lightning	3	318

The Gun Point data [RaK04] are recorded from the video surveillance study and applied in classification tasks. Two types of patterns and some randomly generated subsequences are included in this data. The two patterns, gun-draw and gun-point, are similar to each other but have different lengths and proportions and are also altered by squeeze mappings. The Gun-draw and gun-point pattern are considered as one motif type to examine the performance of KITE. The 50 Words data set [RaM07] consists of 25 motifs of word profiles from George Washington's manuscripts. Food data are applied in chemometrics to classify various sorts of foods. This data set is obtained from the spectrograph of the various sorts of coffee and olive oils [BDH$^+$12]. The Lightning data set [EHD$^+$02] has seven types of motifs that demonstrate similar events associated with lightning. However, to analyse KITE's performance regarding the discovery of ill-known motifs, the author categorises these seven patterns into three groups.

6.2.1.2 Real-World Data

Two data sets, AutoSense-1 and AutoSense-2, are gathered from the research project "Adaptive energy self-sufficient sensor network for monitoring safety-critical self-service-systems" [DLW+15, TDDL16]. This project focuses on monitoring security critical-systems, e.g. identifying criminal attacks on Automated Teller Machines (ATMs). These data sets are obtained through two approaches.

AutoSense-1: In order to observe the physical state of an ATM, a component is equipped with two piezoelectrical sensors (S1 and S2) and one actuator. The actuator enables the component to vibrate, and the sensor converts the vibrations into a voltage signal. The actuator and sensors are connected to a measuring system, which collects the voltage generated by the sensors and controls the actuator simultaneously [TDDL16]. The actuator is intended to stimulate the component's eigenfrequencies because structural modifications to the component change them [DLW+15]. In order to find suitable eigenfrequencies, a FEM-based (Finite Element Method) modal analysis was performed [KST+14]. For the stimulation of the component's eigenfrequencies, a sweep-signal is applied [TDDL16]. A sweep is an alternating voltage of constant amplitude, whose frequency periodically and continuously passes through a specified frequency range [HKV08].

To simulate a physical attack or to skim an ATM, structural manipulations to the component were performed. A graphical representation of a signal $x[n]$ recorded from the sensor S1 during various manipulations is given in Fig. 6.1 (a).

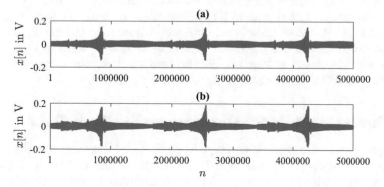

Figure 6.1: AutoSense-1; (a) Signal $x[n]$ is gathered from sensor S1 during various structural manipulations. (b) A non-manipulated signal gathered from sensor S1.

AutoSense-2: For this data set, nine piezoelectrical sensors (S1 to S8 and A1) are employed on a polycarbonate plate. These sensors record the physical alteration of the system. The sensors S1 to S8 are considered passive, and A1 is active, which operates as an actuator. Similar to AutoSense-1, in order to stimulate eigenfrequencies of the component, a sweep-signal is applied. AutoSense-2 data are gathered during three experiments. The component is manipulated by drilling three holes of various diameters. In each test, the sensors S1-S8 gather the state of the component. Figure 6.2 depicts the data obtained from sensor S1 within three physical attempts.

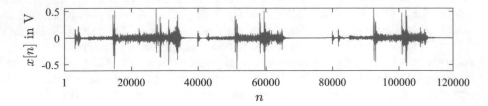

Figure 6.2: Signal $x[n]$ obtained from sensor S1 during three experiments. In each experiment, a hole with different diameters is drilled into the component.

6.3 Detection of Equal-Length Motifs

The detected motifs of equal length from both synthetic and real-world data are presented in this section. In the case of synthetic data, the length and types of motifs are provided in Table 6.2. All of the signals are divided into subsequences of equal lengths by applying Algorithm 1 and are forwarded to KITE's next steps for motif discovery. The results of KITE are validated with the quality measures described in Sec. 6.1.2 and are benchmarked with state-of-the-art algorithms. It should be noted that, since the motif types are known in these experiments, the data sets are considered as labelled. In the next sections, the outcomes obtained in each step of KITE for the stated test case are explained.

6.3.1 Equal-Length Motif Discovery on Synthesis Data

KITE's performance regarding the equal-length motif discovery on the synthesis data sets is evaluated in this section. The results of the Lightning data set are provided in Appendix 11 since its performance is similar to the 50 Words data set.

Gun Point Data Set

The Gun Point data are segmented into subsequences with a lengths of 150 sample by Algorithm 1 in the pre-processing step. In the representation step, each subsequence is decomposed by the ACQTWP transformation, up to a selected scale provided by the early abounding algorithm. This particular scale is mostly the first or second scale for the Gun Point data. For each subsequence, features are extracted from one complex approximation and one complex detail node selected by the Best Nodes Selection (BNS) algorithm. The outcomes of the pre-processing step, early abandoning, and BNS algorithms are given in Table 6.3.

Table 6.3: Results of the pre-processing and representation steps: the number of
$N_s = 60$ subsequences with a length of 150 samples are obtained. The
early abandoning algorithm results in $s_b \in \{1, 2\}$. The BNS algorithm
eventuates the following complex approximation nodes: ${}^sQ_{\mathbb{C},1}$, ${}^sQ_{\mathbb{C},2}$ or
${}^sQ_{\mathbb{C},5}$, and the complex detail nodes ${}^sR_{\mathbb{C},3}$, ${}^sR_{\mathbb{C},4}$, ${}^sR_{\mathbb{C},8}$ or ${}^sR_{\mathbb{C},16}$.

	Outcomes					
Pre-processing	s_1	s_2	s_3	s_4	...	s_{N_s}
Early abandoning	1	1	2	1	...	2
${}^sBQ_{\mathbb{C},i}$	1	2	2	1	...	5
${}^sBR_{\mathbb{C},i}$	4	3	8	3	...	4

LDA evaluates the efficiency of the extracted features, and for all feature combinations, it results in the smallest error of $e = 0$ and the maximum error of $e = 0.3$
when three features are applied.

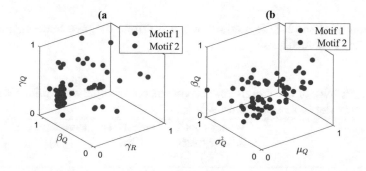

Figure 6.3: Extracted features of motif types 1 and 2. (a) The skewness and kurtosis of the selected complex approximation and detail nodes, (b) The
mean value, variance and kurtosis of selected complex detail and approximation nodes.

Fig. 6.3 illustrates the feature space for two motif types 1 and 2, employing three
different features extracted from the two selected complex approximation and detail
nodes. Motif type 1 contains the randomly generated subsequences, and the actual
patterns in the Gun data set are included in motif type 2. As demonstrated, these
two motif types can be distinguished. Nevertheless, for some feature combinations,
the distance between the two clusters is small. But they are still separable. The
extracted features are sent to a distance measure in order to detect motifs of equal
length. The results of six distance measures are given in Table 6.4.
Since most of the feature combinations result in a classification error less than 0.3,
most of the motifs are identified by the tested distance measures, as confirmed in
Table 6.4. DTW achieved the highest CR, and the Edit distance obtains the lowest
CR.

Table 6.4: Evaluation of KITE equal-length motif discovery and six distance measures on the Gun Point data.

Distance Measure	Correct motif discovery rate (CR)	Sensitivity (Sn)	Precision (Pr)	F-Measure (F-M)
Euclidean Distance (ED)	0.87	0.87	0.86	0.87
Canberra Distance (CD)	0.90	0.98	0.87	0.92
Dynamic Time Warping (DTW)	1.00	1.00	1.00	1.00
Edit Distance	0.75	0.75	0.73	0.74
Edit Distance on Real Sequence (EDR)	0.74	0.75	0.74	0.74
Longest Common SubSequence (LCSS)	0.79	0.80	0.79	0.79

The outcomes of the KITE method, given in Table 6.5, are benchmarked against state-of-the-art algorithms.

Table 6.5: Results of the equal-length motif discovery applying different algorithms for the Gun Point data set.

Method	Correct motif discovery rate (CR)	Sensitivity (Sn)	Precision (Pr)	F-Measure (F-M)
KITE	1.00	1.00	1.00	1.00
Brute-Force [PKL+02]	0.88	0.88	0.88	0.88
MOEN Enum. [MuC15]	0.90	0.90	1.00	0.95
MOEN [MuK10]	0.76	0.76	1.00	0.86
Mr.Motif [CaA10]	1.00	1.00	1.00	1.00
SCRIMP++ [ZYZ+18]	0.81	0.81	0.97	0.88
VALMOD [LZP+18]	0.61	0.61	1.00	0.76

Besides KITE, Mr.Motif [CaA10] also provides the highest CR. Mr.Motif is also able to detect motifs transformed by squeeze and uniform scaling mapping, as long as these motifs have the same length. The maximum amount of precision rate is obtained by KITE, MOEN [MuK10], MOEN Enum. [MuC15], Mr.Motif [CaA10], and VALMOD [LZP+18] algorithms. Discovering ill-known motifs that are stretched or altered by squeeze mapping is challenging for SCRIMP++ and VALMOD, and these methods did not identify such motifs.

The first-representative motif pair obtained by KITE is depicted in Fig. 6.4. Among the detected motifs, this motif pair contains the two most similar subsequences.

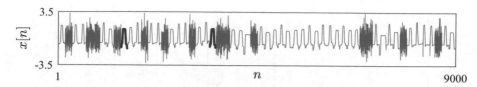

Figure 6.4: First-representative motif pair. Both subsequences have equal length and belong to the motif type gun draw.

50 Words Data Set

A pre-defined length of 270 samples is assigned to motifs of the 50 Words data set. All the subsequences are analysed by the ACQTWP transform after segmentation in the pre-processing step. The results of the early abandoning and BNS algorithms are provided in Table 6.6.

Table 6.6: Results of the pre-processing and representation steps: pre-processing generates $N_\mathrm{s} = 291$ subsequences with a length of 270 samples. The early abandoning algorithm results in $s_\mathrm{b} \in \{1, 2\}$. The BNS algorithm eventuates the following complex approximation nodes: ${}^s Q_{\mathbb{C},1}$, ${}^s Q_{\mathbb{C},2}$, ${}^s Q_{\mathbb{C},5}$, or ${}^s Q_{\mathbb{C},6}$ and the complex detail nodes ${}^s R_{\mathbb{C},3}$, ${}^s R_{\mathbb{C},4}$, ${}^s R_{\mathbb{C},7}$, ${}^s R_{\mathbb{C},8}$ or ${}^s R_{\mathbb{C},12}$.

	Outcomes					
Pre-processing	s_1	s_2	s_3	s_4	...	s_{N_s}
Early abandoning	1	1	2	1	...	2
${}^s BQ_{\mathbb{C},i}$	1	2	2	5	...	6
${}^s BR_{\mathbb{C},i}$	4	3	8	12	...	7

The performance of the feature extraction step is analysed by the classification error obtained from the LDA algorithm. This error is equal to $0.04 \le e \le 0.45$. Employing only two features results in $e = 0.45$, and $e = 0.04$ is obtained when more than five features are utilised.

These features do not attain the least classification error, $e = 0$, and therefore, motifs of the 50 Words data cannot be separated linearly. This issue is illustrated in Fig. 6.5 (a), where the distance between features of the motif types 1 and 4 is not enough to be linearly separated. Additionally, some motif types of the 50 word data cannot be identified, as depicted in Fig. 6.5 (b) and (c). There are subsequences of the motif types 8 and 5 that are considered outliers and cannot be classified correctly by the LDA method. For these groups, the extracted features do not represent the main characteristic of the data.

The reason for this is that the motif types of the 50 Words data set are similar to each other. As an example, the motif type 1 is partly identical to the motif type 2, as illustrated in Fig. 6.6. Therefore, the extracted features are settled very closely to each other in the feature space, making the motif detection challenging.

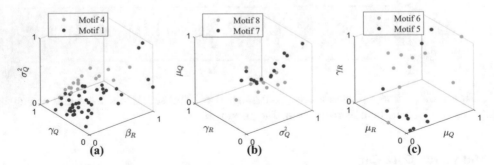

Figure 6.5: Feature space of motifs in the 50 Words data considering three features combination. (a) Motif types 1 and 4 with features σ_Q^2, γ_Q and β_R; (b) Motif types 7 and 8 with features μ_Q, γ_R and σ_Q^2; (c) Motif types 5 and 6 with features γ_R, μ_R and μ_Q.

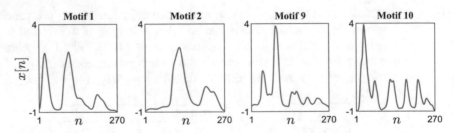

Figure 6.6: Motifs types 1, 2, 9, and 10 of the 50 Words data. The motif types 1 and 2 are similar to each other, likewise the motif types 9 and 10.

Nevertheless, the performance of KITE applying six distance measures is given in Table 6.7. Overall, the tested distance measures are executed akin to each other. The highest CR is achieved by EDR (57%), and the LCSS method obtains the lowest CR (32%). Hence, KITE provides unsatisfactory results (around 50%) for the 50 Words data. As already explained, this is mainly due to the considerable similarity between the given motif types and the extracted features. For example, KITE considers motif type 9 as the ill-known version of motif type 10 due to their amplitude differences.

The performance of KITE is benchmarked against other algorithms in Table 6.8. Although the result of KITE, Mr.Motif [CaA10], and Brute-Force [PKL+02] are similar, Mr.Motif [CaA10] receives the best performance. This shows that for the 50 word data, a symbolic representation, like SAX [LKW+07], results in a higher CR than other methods. belongs to the top-two motifs. SCRIMP++ [ZYZ+18] and VALMOD [LZP+18] achieved the lowest CR. These algorithms as well as MOEN Enum. [MuC15] and Brute-Force [PKL+02] perform directly on the data without any transformation or representation.

Table 6.7: Results of the equal-length motifs in the 50 Words data applying various distance measures.

Distance Measure	Correct motif discovery rate (CR)	Sensitivity (Sn)	Precision (Pr)	F-Measure (F-M)
Euclidean Distance (ED)	0.45	0.45	0.45	0.45
Canberra Distance (CD)	0.48	0.48	0.48	0.48
Dynamic Time Warping (DTW)	0.45	0.45	0.45	0.45
Edit Distance	0.45	0.44	0.43	0.45
Edit Distance on Real Sequence (EDR)	0.57	0.56	0.57	0.57
Longest Common SubSequence (LCSS)	0.32	0.32	0.31	0.31

Table 6.8: Equal-length motif discovery applying different algorithms for the 50 Words data set.

Method	Correct motif discovery rate (CR)	Sensitivity (Sn)	Precision (Pr)	F-Measure (F-M)
KITE	0.57	0.57	0.57	0.57
Brute-Force [PKL+02]	0.58	0.58	0.58	0.58
MOEN Enum. [MuC15]	0.56	0.56	0.57	0.56
MOEN [MuK10]	0.52	0.52	0.51	0.52
Mr.Motif [CaA10]	0.60	0.60	0.58	0.59
SCRIMP++ [ZYZ+18]	0.43	0.43	0.51	0.47
VALMOD [LZP+18]	0.44	0.44	0.53	0.48

Overall, as depicted in Tables 6.7 and 6.8, the correct motif discovery is under 60 % due to the explained reasons (resemblance between the given motif types and poor features for KITE).

Food Data Set

The Food data set has two motif types known to the user: coffee (Motif Type 1) and olive oil (Motif Type 2). The subsequences of these motif types are altered by transformations such as amplitude translation and uniform scaling mapping. Moreover, some subsequences are covered with noise, as depicted in Fig. 6.7.

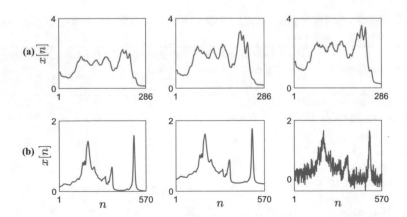

Figure 6.7: Motif types of the Food data; (a) coffee subsequences with a length of
286 samples, (b) olive oil subsequences with a length of 570 samples.

KITE begins by segmenting the Food data into subsequences of 286 samples. Next,
all of the subsequences are decomposed by ACQTWP in the representation step.
The outcomes of this step are given in Table 6.9.

Table 6.9: Results of the pre-processing and representation steps: pre-processing
generates the number of $N_\mathrm{s} = 70$ subsequences with a length of 286
samples. The early abandoning algorithm results in $s_\mathrm{b} \in \{1, 2\}$, and the
BNS algorithm eventuates the following complex approximation nodes:
$^sQ_{\mathbb{C},1}$, $^sQ_{\mathbb{C},2}$, or $^sQ_{\mathbb{C},6}$ and the complex detail nodes $^sR_{\mathbb{C},3}$, $^sR_{\mathbb{C},4}$, $^sR_{\mathbb{C},7}$,
or $^sR_{\mathbb{C},8}$.

	Outcomes					
Pre-processing	s_1	s_2	s_3	s_4	...	s_{N_s}
Early abandoning	1	2	2	1	...	2
$^sBQ_{\mathbb{C},i}$	1	2	2	6	...	6
$^sBR_{\mathbb{C},i}$	4	3	8	4	...	7

For each subsequence, features are extracted from the selected complex nodes. The
classification error of LDA applying these features is $0.01 \leq e \leq 0.06$. Despite the
small amount of classification error, the motif types cannot be completely separated by LDA. As presented in each sub-figure of Fig. 6.7, there are subsequences
considered outliers and assigned to the wrong motif type.

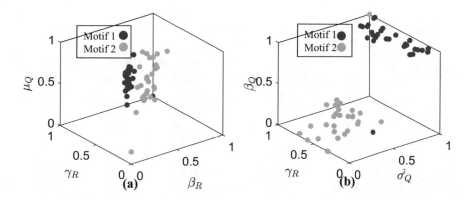

Figure 6.8: Feature space of motifs in the Food data considering three feature values. (a) motif types 1 and 2, and features μ_Q, γ_R and β_R; (b) features β_Q, γ_R, and σ_Q^2 for the same motif types.

According to KITE's structure, motifs are detected by measuring the similarity between features of segmented subsequences. The performance of the distance measures under investigation is presented in Table 6.10. Among the tested distance measures, the EDR method outperformed the other measures due to its robustness to noise [COO05]. The ED and DTW achieved the lowest CR since they are sensitive to outliers.

The efficiency of KITE is benchmarked with other related algorithms in Table 6.11. The highest CR is obtained by KITE and Mr.Motif [CaA10] algorithms (95 %). Nonetheless, the amount of the false-negative rate of Mr.Motif [CaA10] is lower than the KITE method. KITE inaccurately identified three of the olive oil motifs as coffee motifs. SCRIMP++ [ZYZ+18] also obtains a higher precision rate than KITE, but the identified motifs by this method are less than KITE.

Table 6.10: Results of the equal-length motifs in the Food data applying various distance measures.

Distance Measure	Correct motif discovery rate (CR)	Sensitivity (Sn)	Precision (Pr)	F-Measure (F-M)
Euclidean Distance (ED)	0.78	0.78	0.76	0.77
Canberra Distance (CD)	0.94	0.94	0.92	0.93
Dynamic Time Warping (DTW)	0.78	0.78	0.77	0.78
Edit Distance	0.92	0.92	0.90	0.91
Edit Distance on Real Sequence (EDR)	0.95	0.95	0.93	0.94
Longest Common SubSequence (LCSS)	0.88	0.88	0.87	0.86

Table 6.11: Comparison of the equal-length motif discovery for the Food data set
            considering different algorithms.

Method	Correct motif discovery rate (CR)	Sensitivity (Sn)	Precision (Pr)	F-Measure (F-M)
KITE	0.95	0.95	0.93	0.94
Brute-Force [PKL+02]	0.95	0.95	0.93	0.94
MOEN Enum. [MuC15]	0.47	0.47	0.92	0.52
MOEN [MuK10]	0.43	0.43	0.81	0.57
Mr.Motif [CaA10]	0.95	0.95	0.98	0.97
SCRIMP++ [ZYZ+18]	0.67	0.67	0.98	0.79
VALMOD [LZP+18]	0.65	0.65	0.93	0.77

The precision rate of VALMOD [LZP+18] is equal to KITE, but this method
achieves a lower CR. The performance of VALMOD [LZP+18] and SCRIMP++
[ZYZ+18] is similar since they employ the same concept. The lowest performance
is gained by MOEN (43 %). As it is similar to MOEN Enum. [MuC15], it applies
the ED measure, which is sensitive to noise.

6.3.2 Equal-Length Motif Discovery on Real-World Data

The results of the equal-length motif discovery of KITE and other state-of-the-art
algorithms on real-world data are explained in the following sections.

AutoSense-1 Data Set

KITE searches for motifs in AutoSense-1 that can be considered as prototype patterns for manipulated and non-manipulated data. Consequently, KITE divides the
input signal into equal-length subsequences. Based on the expert's knowledge, motifs length must be set to $l_d = 5 \cdot 10^5$. A number of 306 subsequences are obtained.
The ACQTWP transform decomposes a subsequence with a length of $5 \cdot 10^5$ up to
19 scales. Nonetheless, the number of decomposed scales is reduced by the early
abandoning algorithm. Table 6.12 shows the results of KITE's representation step.
Afterwards, the performance of the extracted features is tested by the LDA method,
which results in an error of $0 \le e \le 0.16$. This lower number of errors indicates a
linear and correct classification for the investigated signals; however, the minimum
error is not achieved by all feature combinations. Thus, not all the corresponding
features can be separated linearly by the LDA method (cf. Fig. 6.9).
The motif types 3 and 5 cannot be divided correctly, but the motif types 1 and
2 also the motif types 2 and 4 are separated correctly. The features of the motif
types 1 and 2 are not very far from each other, as depicted in Fig. 6.9 (a), but
the distance between them is large enough to separate them linearly and correctly.
Not all the subsequences of the motif types 3 and 5 can be assigned correctly,
considering the extracted features.

Table 6.12: AutoSense-1 results of the pre-processing and representation steps: in pre-processing, $N_s = 306$ subsequences with a length of 50000 samples are generated. The early abandoning algorithm results in $s_b \in \{5, 6, 7\}$. The best complex approximation nodes, selected by the BNS method, are ${}^sQ_{\mathbb{C},1}$, ${}^sQ_{\mathbb{C},2}$, ${}^sQ_{\mathbb{C},5}$ or ${}^sQ_{\mathbb{C},6}$. The complex detail nodes are ${}^sR_{\mathbb{C},3}$, ${}^sR_{\mathbb{C},12}$, ${}^sR_{\mathbb{C},17}$, ${}^sR_{\mathbb{C},69}$, or ${}^sR_{\mathbb{C},82}$.

	Outcomes					
Pre-processing	s_1	s_2	s_3	s_4	...	s_{N_s}
Early abandoning	2	2	2	2	...	2
${}^sBQ_{\mathbb{C},i}$	1	2	2	5	...	6
${}^sBR_{\mathbb{C},i}$	4	3	7	3	...	8

One subsequence of the motif type 3 is mapped incorrectly in the feature space, and it is classified as motif type 5. Additionally, two other subsequences of the motif type 3 cannot be correctly separated (cf. sub-figure (c)). The same problem occurs for the features of motif types 1 and 4, which, for clarity of the image, is not depicted in Fig 6.9.

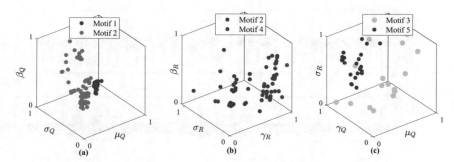

Figure 6.9: Features of five motif types extracted from AutoSense-1 data. (a) motif types 1 and 2 are separated applying μ_Q, σ_Q and β_Q. (b) motif types 2 and 4 are divided correctly and linearly by three features γ_R, σ_R and β_R. (c) motif types 3 and 5 cannot be distinguished correctly.

KITE's performance for AutoSense-1 and the tested distance measures is given below. The CD method obtains the highest CR, and the ED and DTW achieve the lowest CR since they are generally more sensitive to noise. The outcomes gained by the Edit-based distance measures are similar.

Table 6.13: KITE equal-length motif discovery for AutoSense-1 and six distance
measures.

Distance Measure	Correct motif discovery rate (CR)	Sensitivity (Sn)	Precision (Pr)	F-Measure (F-M)
Euclidean Distance (ED)	0.52	0.52	0.51	0.51
Canberra Distance (CD)	0.84	0.84	0.82	0.83
Dynamic Time Warping (DTW)	0.52	0.52	0.51	0.51
Edit Distance	0.68	0.68	0.66	0.67
Edit Distance on Real Sequence (EDR)	0.69	0.69	0.67	0.68
Longest Common SubSequence (LCSS)	0.63	0.63	0.62	0.62

However, the EDR measure handles outliers better than other methods such as
the ED or DTW and obtains more accurate results than LCSS [COO05]. KITE
detects five types of motifs. The equivalence class of first-frequent motifs $[^1m]$
contains signals that resemble events without physical manipulation, as depicted
in Fig. 6.10.

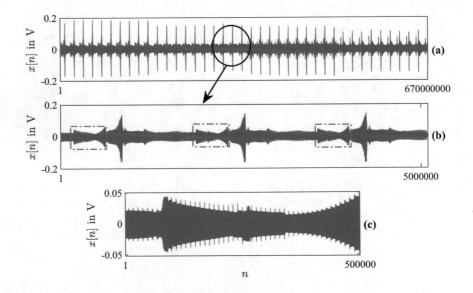

Figure 6.10: Sub-figure (a) represents parts of the AutoSense-1 data set, and in (b)
a zoom view from a section of the AutoSense-1 data is given. Sub-
figure (c) depicts the first-representative motif obtained by KITE's
equal-length motif discovery.

The obtained motif type (depicted in Fig. 6.10 (c)) appears mostly at the beginning of the non-manipulated signals. Therefore, in order to detect an anomaly or classify signals, the beginning of the signal can be examined instead of analysing the entire signal. This lowers the computational time for the classification task. Moreover, this is a critical issue in on-line applications where the data are streaming [TDDL16].

KITE's efficiency is compared with state-of-the-art motif discovery algorithms, provided in Table 6.14. KITE outperformed other methods by achieving the CR of 84%. The CR achieved by SCRIMP++ [ZYZ+18] is identical to KITE, but KITE's precision is higher than SCRIMP++. Mr.Motif [CaA10] obtained a correct motif discovery rate of 62%. The Brute-Force [PKL+02] method cannot detect all of the motifs and results in a correct motif discovery rate of 26%. The results of this evaluation confirm that KITE's representation and feature extraction steps lead to higher correct motif discovery and precision rate than other methods. SCRIMP++ [ZYZ+18] as well as the Brute-Force [PKL+02] methods analyse the data in the time domain without employing any representation methods. The lower motif discovery of Mr.Motif [CaA10] indicates that the performance of SAX is sensitive to noise. The VALMOD [LZP+18], MOEN Enum. [MuC15], and MOEN [MuK10] methods are not able to execute the motif discovery because of the size of AutoSense-1 data.

Table 6.14: Equal-length motif discovery applying various algorithms for the AutoSense-1 data set.

Method	Correct motif discovery rate (CR)	Sensitivity (Sn)	Precision (Pr)	F-Measure (F-M)
KITE	0.84	0.84	0.82	0.83
Brute-Force [PKL+02]	0.26	0.26	0.25	0.25
MOEN Enum. [MuC15]	-	-	-	-
MOEN [MuK10]	-	-	-	-
Mr.Motif [CaA10]	0.62	0.62	0.61	0.61
SCRIMP++ [ZYZ+18]	0.83	0.83	0.66	0.62
VALMOD [LZP+18]	-	-	-	-

AutoSense-2 Data Set

The number and lengths of motifs in AutoSense-2 data are not provided in advance. Consequently, for the first evaluation, signals of the same sensor and its adjacent sensor are considered motifs, as recommended by the expert. Signals of the AutoSense-2 data are segmented into subsequences of 39000 samples. Without the early abandoning algorithm, the ACQTWP transform must decompose each subsequence up to $s_T = 15$ scales. However, by executing the early abandoning algorithm, the decomposition stops before s_T, and the entire scales must not be analysed. The results of the representation step are given in Table 6.15.

Table 6.15: Results of the pre-processing and representation steps for AutoSense-2:
pre-processing generates $N_s = 24$ subsequences with a length of 39000
samples. The early abandoning algorithm results in $s_b \in \{5, 6, 7\}$. The
best complex approximation nodes selected by the BNS algorithm are
$^sQ_{C,1}$, $^sQ_{C,2}$, $^sQ_{C,5}$ or $^sQ_{C,6}$. The complex detail nodes are $^sR_{C,3}$,
$^sR_{C,12}$, $^sR_{C,17}$, $^sR_{C,69}$ or $^sR_{C,82}$.

	Outcomes					
	s_1	s_2	s_3	s_4	...	s_{N_s}
Pre-processing						
Early abandoning	7	5	7	7	...	6
$^sBQ_{C,i}$	1	2	2	5	...	6
$^sBR_{C,i}$	69	3	17	21	...	82

After extracting features from the selected nodes' coefficients, their efficiency is
examined by the LDA method. The features of four motif types are depicted in
Fig. 6.11.

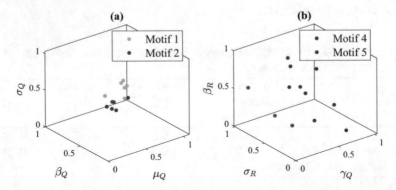

Figure 6.11: Features of 4 motif types extracted from the AutoSense-2 data. (a)
the motif types 1 and 2 presented by three features μ_Q, β_Q, and σ_Q;
(b) the motif types 4 and 5 depicted by three features γ_Q, σ_R, and β_R.

The motif types 4 and 5, illustrated in sub-figure (b), can be separated linearly
by the LDA method. However, the motif types 1 and 2 cannot be divided linearly
by the LDA method since one subsequence of each motif type cannot be classified
correctly (cf. sub-figure (a)).

The classification error obtained by LDA from the aforementioned features is $0 \le
e \le 0.16$.

Subsequently, a distance measure quantifies the similarity between the derived
features to determine motifs. The performance of the six tested distance measures
is presented in Table 6.16. The highest CR (61 %) is obtained by the EDR, and the
ED results in the lowest rate of 49 %. Among the tested measures, the EDR is more

robust to noise than others and therefore obtains the highest CR. The Canberra, Edit, and LCSS distances obtained a correct motif discovery of the same range.

Like the rest of the test cases, KITE's performance is compared with the aforementioned methods in Table 6.17. KITE provides the highest CR in comparison with other state-of-the-art algorithms. Apart form KITE and Mr.Motif [CaA10], other evaluated methods apply the ED method in their similarity measurement step and therefore obtain poor results.

The MOEN [MuK10] algorithm detects motifs if the similar subsequences are directly located after each other or if they are in the same sliding window. This method is applied for on-line motif discovery and therefore can only compare the recent subsequence with the exact one before. The CR of VALMOD [LZP$^+$18] is identical with the MOEN [MuK10] method. Although not all the motifs are identified by VALMOD [LZP$^+$18], it achieved the highest precision rate. The CR achieved by SCRIMP++ [ZYZ$^+$18] is similar to KITE, but the precision rate of SCRIMP++ [ZYZ$^+$18] is higher than KITE.

Table 6.16: Results of the equal-length motifs in the AutoSense-2 data applying six distance measures.

Distance Measure	Correct motif discovery rate (CR)	Sensitivity (Sn)	Precision (Pr)	F-Measure (F-M)
Euclidean Distance (ED)	0.49	0.49	0.50	0.49
Canberra Distance (CD)	0.56	0.56	0.55	0.56
Dynamic Time Warping (DTW)	0.50	0.50	0.50	0.50
Edit Distance	0.58	0.58	0.58	0.58
Edit Distance on Real Sequence (EDR)	0.61	0.61	0.60	0.60
Longest Common SubSequence (LCSS)	0.54	0.54	0.51	0.52

In order to improve KITE's performance, it is possible to extract and add other features. As stated earlier, one of KITE's characteristics is the ability to add or omit features. Thus, in addition to the six applied features, two other features, namely energy density of the wavelet coefficients (cf. Def. 5.16) and root mean square [Tho65] are considered. These features belong to the commonly-applied features employed for vibration signals [BDM$^+$12, REJ$^+$17, FJH$^+$18, YKA$^+$19].

Table 6.18 depicts the outcomes of KITE employing the eight mentioned features and the tested distance measures.

By employing the two recent features, the highest CR (78 %) is obtained by the EDR measure. The lowest CR is achieved by the ED. These former outcomes are not considered in KITE's evaluation, as the extracted features must be the same for all test cases and methods. However, this example indicates that if the proposed

Table 6.17: Equal-length motif discovery applying various algorithms for the AutoSense-2 data set.

Method	Correct motif discovery rate (CR)	Sensitivity (Sn)	Precision (Pr)	F-Measure (F-M)
KITE	0.61	0.61	0.60	0.60
Brute-Force [PKL+02]	0.47	0.47	0.48	0.47
MOEN Enum. [MuC15]	0.45	0.45	0.91	0.61
MOEN [MuK10]	0.34	0.34	0.32	0.33
Mr.Motif [CaA10]	0.48	0.48	0.46	0.46
SCRIMP++ [ZYZ+18]	0.58	0.58	0.66	0.62
VALMOD [LZP+18]	0.33	0.33	1.00	0.50

Table 6.18: Results of the equal-length motifs in the AutoSense-2 data applying six distance measures.

Distance Measure	Correct motif discovery rate (CR)	Sensitivity (Sn)	Precision (Pr)	F-Measure (F-M)
Euclidean Distance (ED)	0.56	0.56	0.56	0.56
Canberra Distance (CD)	0.65	0.69	0.70	0.71
Dynamic Time Warping (DTW)	0.56	0.56	0.56	0.56
Edit Distance	0.65	0.65	0.60	0.61
Edit Distance on Real Sequence (EDR)	0.78	0.78	0.75	0.76
Longest Common SubSequence (LCSS)	0.58	0.58	0.51	0.51

features cannot represent the characteristics of the data, KITE's performance can be improved by including or excluding features.

6.3.3 Equal-Length Motif Discovery Summary

A review of KITE's equal-length motif discovery is provided in Table 6.19. KITE's performance is mostly superior or equal to state-of-the-art algorithms, except for the 50 words data set. As explained in Sec. 6.3.1, the motif types of the 50 words data are very similar. As an example, despite the differences in amplitude and proportion, subsequences of the motif types 9 and 10 resemble each other. Consequently, these subsequences are regarded as ill-known motifs and are determined as one motif type.

Among the investigated methods, MOEN Enum. [MuC15], MOEN [MuK10], and VALMOD [LZP+18] collapse when the size of the data and the motif's length are

Table 6.19: Summary of equal-length motif discovery. CR denotes the correct motif discovery rate is denoted. KITE's employed distance measure for the highest CR is provided. Euclidean distance (ED), Canberra distance (CD), Edit Distance on Real Sequence (EDR).

Data	Highest CR	Method	Lowest CR	Method
Gun Point	0.90	KITE (CD), Mr.Motif	0.61	VALMOD
50 Words	0.60	Mr.Motif	0.43	SCRIMP++
Food	0.95	KITE (EDR), Mr.Motif Brute Force	0.43	MOEN
Lightning	0.77	KITE (EDR), Mr.Motif	0.44	MOEN
AutoSense-1	0.84	KITE (CD)	0.26	Brute-Force
AutoSense-2	0.61	KITE (EDR)	0.33	VALMOD

large ($N > 10^6$). The performance of Mr.Motif [CaA10] and KITE for most of the test cases was similar. A reason is that Mr.Motif [CaA10] and KITE are the only methods among the tested ones that apply a representation method. Moreover, all the evaluated methods measure the similarities by the ED method. This is challenging for test cases like AutoSense 1 and 2, where the data is superimposed with noise since the ED is noise sensitive. Between the tested distance measures, the EDR and CD obtained the highest CR for most of the test cases. This is due to the robustness of EDR to noise compared with other methods such as the ED or DTW [COO05]. LCSS is the most sensitive measure to noise among the tested methods, which leads to poor results.

The outcome of the KITE algorithm shows that the six stated features (the four first statistical moments and the maximum and minimum phase of the coefficients) are mostly enough to detect motifs. Nevertheless, as illustrated in Sec. 6.3.2, the correct motif discovery can be improved by including or excluding other features.

6.4 Detection of Variable-Length Motifs

The performance of KITE's variable-length motif discovery on both synthetic and
real-world data is presented in this section. The results of the KITE variable-length
motif discovery are evaluated with the quality measures described in Sec. 6.1.2 and
are compared with the outcomes of the MOEN Enum. [MuC15] and VALMOD
[LZP+18] methods. Among state-of-the-art algorithms, only these two methods
discover motifs of variable lengths in the same way as KITE. A comprehensive
explanation of the obtained results is given in the next sections.

6.4.1 Variable-Length Motif Discovery on Synthesis Data

This section evaluates KITE's variable-length motif discovery on the synthesis data.
Similar to the equal-length motif discovery, the variable-length motif discovery
results on the Lightning data set are provided in Appendix 11.

Gun Point Data Set

The Gun Point data set is divided into 158 subsequences of lengths between 110
to 150 samples by algorithm 2. Table 6.20 represents the results obtained from the
representation step of KITE.

Table 6.20: Results of the pre-processing step, early abandoning, and BNS algorithms obtained from the Gun Point data. The number of $N_s = 158$
subsequences with lengths between 110 to 150 samples are generated in
pre-processing. The early abandoning algorithm results in $s_b \in \{1, 2\}$.
The BNS algorithm eventuates $^sQ_{C,1}$ or $^sQ_{C,2}$ as the best complex
approximation node, and the complex detail node is $^sR_{C,3}$ or $^sR_{C,4}$.

	Outcomes					
	s_1	s_2	s_3	s_4	...	s_{N_s}
Pre-processing						
Early abandoning	1	1	2	1	...	1
$^sBQ_{C,i}$	1	2	1	2	...	1
$^sBR_{C,i}$	4	3	3	3	...	4

The ACQTWP transform decomposes the subsequences mostly up to the third
scale, and the early abandoning algorithm determines $s_b \in \{1, 2\}$. The BNS algorithm selects the two most advantageous nodes for the feature extraction step. The
best detail node is selected from $^sR_{C,3}$ and $^sR_{C,4}$, and the selected approximation
node is specified from $^sQ_{C,1}$ or $^sQ_{C,2}$. Subsequently, the features are extracted
from the normalised coefficients of the stated nodes. As the performance of these
features is comprehensively explained in the previous sections, it is not included
here.

Finally, by measuring the similarity between all of the subsequences, motifs are
detected. The results of KITE variable-length motif discovery for the Gun Point
data and the six distance measures are illustrated in Table 6.21.

Table 6.21: KITE's variable-length motif discovery for the Gun Point data applying six distance measures.

Distance Measure	Correct motif discovery rate (CR)	Sensitivity (Sn)	Precision (Pr)	F-Measure (F-M)
Euclidean Distance (ED)	0.88	0.86	0.87	0.88
Canberra Distance (CD)	0.88	0.86	0.86	0.87
Dynamic Time Warping (DTW)	0.78	0.76	0.77	0.76
Edit Distance	0.60	0.58	0.58	0.59
Edit Distance on Real Sequence (EDR)	0.81	0.81	0.80	0.80
Longest Common SubSequence (LCSS)	0.69	0.68	0.68	0.69

The ED and CD obtain the highest CR (88 %). Nevertheless, if the precision and the F-measure rates are inspected, the ED performs more precisely than CD. The lowest CR is achieved by the Edit distance. The outcome of the KITE method is compared with the result of MOEN Enum. [MuC15] and VALMOD [LZP+18] in Table 6.22. MOEN Enum. [MuC15] and VALMOD [LZP+18] cannot determine all of the motifs, and their performance is less than the KITE method.

Table 6.22: Results of KITE, MOEN Enum. and VALMOD variable-length motifs for the Gun Point data.

Method	Correct motif discovery rate (CR)	Sensitivity (Sn)	Precision (Pr)	F-Measure (F-M)
KITE	0.88	0.86	0.87	0.88
MOEN Enum. [MuC15]	0.62	0.62	1.00	0.57
VALMOD [LZP+18]	0.57	0.57	0.85	0.68

The precision rate of VALMOD [LZP+18] and MOEN Enum. [MuC15] (85 % and 100 %) is higher than KITE. This indicates a higher amount of the true-positive and false-positive rate obtained by these methods. However, the rate for false-negative obtained from these methods is higher than KITE, considering the sensitivity and F-Measure rate. The first-representative motif of the Gun Point data detected by KITE comprises two ill-known motifs with various lengths, and one of them is altered by the stretch mapping, as depicted in Fig. 6.12.

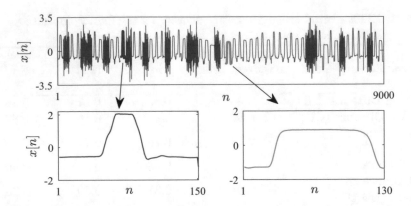

Figure 6.12: The first-representative motif pair obtained by KITE. These two sub-sequences have different lengths, and both belong to the same motif group.

50 Words Data Set

The results of KITE's pre-processing and representation steps for variable-length motif discovery of the 50 Words data set are provided in Table 6.23.

Table 6.23: Results of the pre-processing step, early abandoning, and BNS algorithms obtained from the 50 Words data. The number of $N_s = 979$ subsequences with the lengths between 210 to 270 samples are generated in pre-processing. The early abandoning algorithm results in $s_b \in \{1, 2\}$. The BNS algorithm eventuates $^sQ_{C,1}$ or $^sQ_{C,2}$ as the best complex approximation node, and the complex detail node is either $^sR_{C,3}$ or $^sR_{C,4}$.

	Outcomes					
Pre-processing	s_1	s_2	s_3	s_4	...	s_{N_s}
Early abandoning	1	2	1	1	...	2
$^sBQ_{C,i}$	1	2	1	2	...	1
$^sBR_{C,i}$	4	3	3	3	...	4

The number of 979 subsequences with different lengths from 210 up to 270 samples are obtained in the pre-processing phase. Regardless of their lengths, all of these signals are analysed by the ACQTWP transform. The outcome of the early abandoning method, for the tested signals is $s_b \in \{1, 2\}$. The best complex approximation node is selected between $^sQ_{C,1}$ or $^sQ_{C,2}$, and from the two following nodes, $^sR_{C,3}$ or $^sR_{C,4}$, one is assigned as the best complex detail node. After extracting features from the selected nodes, motifs are detected by employing distance measures, as depicted in Table 6.24.

Table 6.24: KITE's variable-length motif discovery for the 50 Words data applying six distance measures.

Distance Measure	Correct motif discovery rate (CR)	Sensitivity (Sn)	Precision (Pr)	F-Measure (F-M)
Euclidean Distance (ED)	0.44	0.44	0.44	0.44
Canberra Distance (CD)	0.57	0.57	0.56	0.56
Dynamic Time Warping (DTW)	0.61	0.61	0.60	0.61
Edit Distance	0.41	0.41	0.41	0.41
Edit Distance on Real Sequence (EDR)	0.50	0.50	0.50	0.50
Longest Common SubSequence (LCSS)	0.40	0.40	0.40	0.40

The highest and the lowest CR is achieved by DTW and LCSS, respectively. The motif discovery rate gained from other distance measures is around 50 % due to the high intra-class similarity between motif types. Figure 6.13 illustrates the first-representative motif pair identified by KITE.

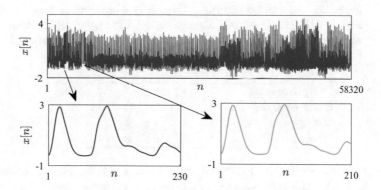

Figure 6.13: 50 Words first-representative motif pair, which is detected by KITE. The two subsequences have different lengths and belong to the same motif type.

As provided in Table 6.25, KITE performs superior to the VALMOD [LZP+18] and MOEN Enum. [MuC15] method. Similar to the equal-length motif discovery, MOEN Enum. [MuC15] cannot determine all motif types. The variable-length motif discovery results of KITE and VALMOD [LZP+18] are slightly better (4-5 %) than their equal-length motif discovery. The reason is that the variable-length motif discovery increases the number of segmented subsequences and the opportunity to identify motifs.

Table 6.25: Results of KITE, MOEN Enum. and VALMOD variable-length motifs
for the 50 Words data.

Method	Correct motif discovery rate (CR)	Sensitivity (Sn)	Precision (Pr)	F-Measure (F-M)
KITE	0.61	0.61	0.60	0.61
MOEN Enum. [MuC15]	0.52	0.52	0.48	0.50
VALMOD [LZP+18]	0.48	0.48	0.46	0.47

Food Data Set

By applying Algorithm 2, the Food data are segmented into 151 subsequences with
various lengths between 300 to 570 samples. As given in the following table, the
signals are decomposed up to the third scale in the representation step.

Table 6.26: Results of the pre-processing step, early abandoning, and Best Nodes
Selection algorithms obtained from the Food data. The number of
$N_s = 151$ subsequences with the lengths between 300 to 570 samples are
generated in pre-processing. The early abandoning algorithm results
in $s_b \in \{1, 2\}$. The BNS algorithm detects $^sQ_{C,1}$ as the best complex
approximation node and the complex detail node is assigned either by
$^sR_{C,3}$ or by $^sR_{C,4}$.

	Outcomes					
	s_1	s_2	s_3	s_4	...	s_{N_s}
Pre-processing						
Early abandoning	1	2	1	1	...	2
$^sBQ_{C,i}$	1	1	1	1	...	1
$^sBR_{C,i}$	4	3	3	3	...	4

The outcome of the early abandoning algorithm is $s_b \in \{1, 2\}$. For all subsequences,
$^sQ_{C,1}$ is selected by the BNS algorithm as the best complex approximation node,
and the best complex detail node is collected between $^sR_{C,3}$ or $^sR_{C,4}$.
Table 6.27 represents the results of KITE employing six distance measures. Among
these distance measures, the ED, CD, and DTW achieve the highest results compared with other measures. The minimum CR (56%) is obtained by the LCSS
method. KITE is more efficient than the MOEN Enum. [MuC15] and VALMOD
[LZP+18] methods regarding variable-length motif discovery. The efficiency is given
in Table 6.28.
KITE obtains a CR of 98%. However, the precision rate is 71%. In contrast to
KITE, VALMOD [LZP+18] and MOEN Enum. [MuC15] achieve a higher precision
rate than KITE (83% and 100%, respectively). Nevertheless, analysing the F-
Measure justifies KITE's performance.

Table 6.27: KITE's variable-length motif discovery for the Food data applying six distance measures.

Distance Measure	Correct motif discovery rate (CR)	Sensitivity (Sn)	Precision (Pr)	F-Measure (F-M)
Euclidean Distance (ED)	0.98	0.98	0.71	0.82
Canberra Distance (CD)	0.97	0.97	0.70	0.81
Dynamic Time Warping (DTW)	0.96	0.96	0.69	0.80
Edit Distance	0.64	0.64	0.61	0.62
Edit Distance on Real Sequence (EDR)	0.70	0.70	0.66	0.68
Longest Common SubSequence (LCSS)	0.56	0.56	0.55	0.56

Table 6.28: Results of KITE, MOEN Enum. and VALMOD variable-length motifs for the Food data.

Method	Correct motif discovery rate (CR)	Sensitivity (Sn)	Precision (Pr)	F-Measure (F-M)
KITE	0.98	0.98	0.71	0.82
MOEN Enum. [MuC15]	0.52	0.52	1.00	0.50
VALMOD [LZP+18]	0.70	0.70	0.83	0.76

The first-representative motif pair of the Food data is illustrated in Fig. 6.14, where both subsequence belong to the same motif type (motif type coffee).

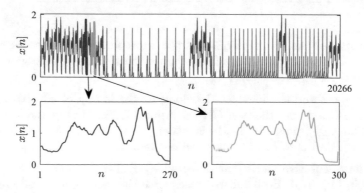

Figure 6.14: The first-representative motif pair of various lengths detected in Food data.

6.4.2 Variable-Length Motif Discovery on Real-World Data

The outcomes of the KITE, MOEN Enum. [MuC15], and VALMOD [LZP+18] methods regarding variable-length motif discovery on real-world data are analysed in this section.

AutoSense-1 Data Set

In order to detect motifs of variable lengths, KITE divides the AutoSense-1 data into subsequences with the lengths ranging from $1 \cdot 10^5$ to $5 \cdot 10^5$ samples applying Algorithm 2. This step results in 456 subsequences. The outcome of the early abandoning algorithm for all of the subsequences is the second scale. Table 6.29 summarises the results of the representation step.

Table 6.29: Results of the pre-processing step, early abandoning, and BNS algorithms obtained from the AutoSense-1 data. The number of $N_s = 456$ subsequences with the lengths between $1 \cdot 10^5$ to $5 \cdot 10^5$ samples are generated in pre-processing. The early abandoning algorithm results in $s_b = 2$. The BNS algorithm detects ${}^s Q_{\mathbb{C},1}$, ${}^s Q_{\mathbb{C},2}$, ${}^s Q_{\mathbb{C},5}$, or ${}^s Q_{\mathbb{C},6}$ as the best complex approximation node and the complex detail node is assigned either by ${}^s R_{\mathbb{C},3}$, ${}^s R_{\mathbb{C},4}$, ${}^s R_{\mathbb{C},7}$, or by ${}^s R_{\mathbb{C},8}$.

	Outcomes					
Pre-processing	s_1	s_2	s_3	s_4	...	s_{N_s}
Early abandoning	2	2	2	2	...	2
${}^s BQ_{\mathbb{C},i}$	1	2	6	1	...	5
${}^s BR_{\mathbb{C},i}$	4	3	3	8	...	7

The similarity between the subsequences is computed by a distance measure. The outcomes of the similarity measurement step for the six under-investigated distance measures are given in Table 6.30.

Both the highest and the lowest CR belong to the family of edit-based distance measures. EDR achieves the best CR (83 %) and LCSS obtains the lowest results (53 %). The performance of LCSS is usually poor due to its noise sensitivity. Nevertheless, EDR is robust to noise and outliers.

The first representative motif pair detected by KITE is similar to the identified motif pair in Fig. 6.10. Both subsequences have the same length and are detected from the signals that resemble normal events (without physical manipulation). The performance of KITE is benchmarked with two other algorithms and is provided in Table 6.31.

Table 6.30: KITE's variable-length motif discovery for the AutoSense-1 data applying six distance measures.

Distance Measure	Correct motif discovery rate (CR)	Sensitivity (Sn)	Precision (Pr)	F-Measure (F-M)
Euclidean Distance (ED)	0.61	0.61	0.79	0.69
Canberra Distance (CD)	0.69	0.69	0.87	0.77
Dynamic Time Warping (DTW)	0.76	0.76	0.88	0.82
Edit Distance	0.63	0.63	0.85	0.71
Edit Distance on Real Sequence (EDR)	0.83	0.83	0.99	0.90
Longest Common SubSequence (LCSS)	0.53	0.53	0.56	0.55

Table 6.31: Results of KITE, MOEN Enum., and VALMOD variable-length motifs for the AutoSense-1 data.

Method	Correct motif discovery rate (CR)	Sensitivity (Sn)	Precision (Pr)	F-Measure (F-M)
KITE	0.83	0.83	0.99	0.90
MOEN Enum. [MuC15]	-	-	-	-
VALMOD [LZP+18]	-	-	-	-

As explained in Sec. 6.3.2, the MOEN Enum. [MuC15] and VALMOD [LZP+18] algorithms are unable to detect motifs because of the size of the data. Thus, the data are divided into smaller segments for the MOEN Enum. [MuC15] and VALMOD [LZP+18] algorithms to determine their performance. The CR obtained from these two methods is 45 % and 57 %, respectively. This is less than KITE analysing the entire data.

AutoSense-2 Data Set

In the pre-processing step, all the signals of the AutoSense-2 are segmented into subsequences with the lengths between 10000 and 40000 samples. Next, these subsequences are analysed by the ACQTWP transform. The outcomes of the early abandoning and BNS algorithms are given in Table 6.32. The efficiency of the third step, feature extraction, has already been explained in Sec. 6.3.2. In the fourth step of KITE, motifs are detected by quantifying their similarity.

The results of the six distance measures under investigation are given in Table 6.33. Like AutoSense-1, in AutoSense-2, the highest CR of 68 % is obtained by EDR and LCSS results in the lowest rate of 52 %.

Table 6.32: Results of the pre-processing step, early abandoning, and BNS algorithms obtained from the AutoSense-2 data. The number of $N_s = 193$ subsequences with the lengths between 10000 to 40000 samples are generated in pre-processing. The early abandoning algorithm results in $s_b \in \{5, 6, 7, 8\}$. The BNS algorithm determines ${}^sQ_{C,1}, {}^sQ_{C,2}, {}^sQ_{C,5}$, or ${}^sQ_{C,6}$ as the best complex approximation node and the complex detail node is assigned either by ${}^sR_{C,3}, {}^sR_{C,4}, {}^sR_{C,7}$, or by ${}^sR_{C,8}$.

	Outcomes					
Pre-processing	s_1	s_2	s_3	s_4	...	s_{N_s}
Early abandoning	5	8	7	8	...	6
${}^sBQ_{C,i}$	1	2	6	5	...	21
${}^sBR_{C,i}$	45	108	60	70	...	20

Table 6.33: KITE's variable-length motif discovery for the AutoSense-2 data applying six distance measures.

Distance Measure	Correct motif discovery rate (CR)	Sensitivity (Sn)	Precision (Pr)	F-Measure (F-M)
Euclidean Distance (ED)	0.54	0.54	0.51	0.52
Canberra Distance (CD)	0.58	0.58	0.58	0.58
Dynamic Time Warping (DTW)	0.56	0.56	0.55	0.56
Edit Distance	0.62	0.62	0.61	0.61
Edit Distance on Real Sequence (EDR)	0.68	0.68	0.66	0.67
Longest Common SubSequence (LCSS)	0.52	0.52	0.51	0.52

The results of KITE are compared with the outcomes of MOEN Enum. [MuC15] and VALMOD [LZP+18] in the following table. Accordingly, KITE outperforms both of the above-stated methods in variable-length motif discovery. MOEN Enum. [MuC15] and VALMOD [LZP+18] obtained the highest precision rate of 100 %, although their CR is less than KITE and their sensitivity is low (less than 50 %). However, analysing the F-Measure indicates that despite the higher amount of the true-positive and false-positive rate obtained by MOEN Enum. [MuC15] and VALMOD [LZP+18], KITE's performance is more accurate.

Finally, the first-representative motif pair discovered among all detected motifs by KITE is given in Fig. 6.15. Both motifs belong to the same motif type, and have variable lengths.

Table 6.34: Results of KITE, MOEN Enum., and VALMOD variable-length motifs
for the AutoSense-2 data.

Method	Correct motif discovery rate (CR)	Sensitivity (Sn)	Precision (Pr)	F-Measure (F-M)
KITE	0.68	0.68	0.66	0.67
MOEN Enum. [MuC15]	0.45	0.45	1.00	0.50
VALMOD [LZP+18]	0.33	0.33	1.00	0.50

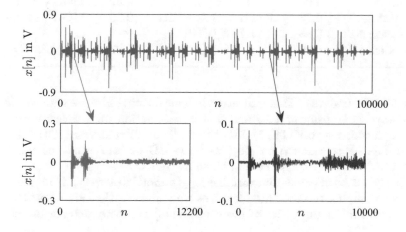

Figure 6.15: The first-representative motif pair of the AutoSense-2 data set detected by KITE.

6.4.3 Variable-Length Motif Discovery Summary

The performance of KITE's variable-length motif discovery is summarised in Table
6.35. The highest CR is mostly obtained by the ED and EDR. For noisy test cases,
EDR performs superior to the ED. The minimum CR for the Gun Point data set is
achieved by the Edit distance, and for the 50 Words, Food, Lightning, AutoSense-1
and 2 data sets, this rate is obtained by LCSS since it is the most sensitive measure
to noise and outliers.

As presented, KITE's outcome is tested against the MOEN Enum. [MuC15] and
VALMOD [LZP+18] methods since these two methods detect variable-lengths motifs without iterating the same algorithm for several defined motifs lengths.

For all the test cases, KITE obtained superior results than MOEN Enum. [MuC15]
and VALMOD [LZP+18] approaches, and this is regardless of the applied distance
measure (excluding 50 Words data). Even in cases where the ED is applied, comparing KITE's results and VALMOD [LZP+18] and MOEN Enum. [MuC15] outcomes
shows that KITE's CR was higher than the two other methods for all test cases (ex-

Table 6.35: Summary of the variable-length motif discovery. The applied distance
measure by KITE is provided in (.), ED: Euclidean distance, CD: Canberra distance, DTW: Dynamic Time Warping, EDR: Edit Distance
on Real Sequence; CR: correct motif discovery rate.

Data	Highest CR	Method	Lowest CR	Method
Gun Point	0.88	KITE (ED & CD)	0.57	VALMOD
50 Words	0.61	KITE (DTW)	0.48	VALMOD
Food	0.98	KITE (ED)	0.52	MOEN Enum.
Lightning	0.73	KITE (ED & EDR)	0.52	MOEN Enum.
AutoSense-1	0.83	KITE (EDR)	-	-
AutoSense-2	0.68	KITE (EDR)	0.33	VALMOD

cept for 50 Words data). This indicates that performing a representation method
such as wavelet transformations and feature extraction approaches improve the
CR. Since in contrast to KITE, MOEN Enum. [MuC15] and VALMOD's [LZP⁺18]
structure has no representation step, and the ED or its normalised version are
employed in the measurement step of these methods.
Additionally, for both equal- and variable-length motif discovery, KITE's outcomes
reveal that applying the six stated features (four first statistical moments and the
maximum and minimum phase of the coefficients) is mostly enough to determine
motifs.

6.5 KITE Robustness Toward Noise

Noise is a typical undesired phenomenon occurring in real-world data and applications. Therefore, the effects of noise on KITE's performance are investigated in this
section. For this reason, the synthetic data sets are covered with noise. The measure to quantify the amount of the noise affecting signal $x[n]$ is the signal-to-noise
ratio (SNR) and is defined by [OpS89, Mad97, Che09]

$$\text{SNR} = 20 \cdot \log \frac{\sigma(\check{N})}{\sigma(x)} \text{dB}, \tag{6.1}$$

where $\check{N}$ is the affecting noise, and $\sigma()$ is the standard deviation.
To obtain unbiased results, all the tests' parameters in this section are equal to the
previous tests in Sec.6.3 and 6.4. Thus, the length of the motifs for both equal- and
variable-length motif discovery does not differ. All the test cases are superimposed
with the Gaussian noise [OpS89] equal to SNR $\in \{30\text{dB}, 20\text{dB}, 10\text{dB}\}$.
Note: The experiments performed with SNR $\in \{30\text{dB}, 20\text{dB}\}$ did not provide significant results. Consequently, KITE's performance in equal-length motif discovery

for data with superimposed noise (SNR=10dB) is provided in Table 6.36.

Table 6.36: KITE's equal-length motif discovery for data with noise and without
noise. CR: correct motif discovery rate. ED: Euclidean distance, CD:
Canberra distance, DTW: Dynamic Time Warping, EDR: Edit Distance on Real Sequence.

	Data with Noise (SNR= 10dB)		Data without Noise	
	Best CR	Distance Measure	Best CR	Distance Measure
Gun Point	0.87	DTW	1.00	DTW
50 Words	0.52	CD	0.57	EDR
Food	0.89	EDR	0.95	EDR
Lightning	0.76	CD, EDR	0.77	CD

By comparing the results obtained from the data with noise and data without noise,
it can be concluded that KITE is robust toward noise. The best CR of KITE for
the noisy data is less than the best CR of the data without noise. Nonetheless,
the difference between these two results is insignificant. Further results of the six
tested distance measures on the noisy data sets are given in Appendix 11.4.

The performance of KITE on noisy data sets is benchmarked against state-of-
the-art algorithms, given in Table 6.37. Except for the 50 Words data, KITE
outperforms other approaches when the data sets are covered with noise. For the
50 word data set, Mr.Motif [CaA10] and SCRIMP++ [ZYZ+18] achieve better
outcomes than KITE. The results of the Brute-Force [PKL+02] method are based

Table 6.37: Equal-length motif discovery on the synthetic data sets covered by noise
(SNR=10dB).

Method/Data	Gun Point	50 Words	Food	Lightning
KITE	0.87	0.52	0.89	0.76
Brute-Force [PKL+02]	0.82	0.50	0.88	0.64
MOEN Enum. [MuC15]	0.84	0.52	0.43	0.50
MOEN [MuK10]	0.75	0.45	0.42	0.41
Mr.Motif [CaA10]	0.53	0.57	0.78	0.76
SCRIMP++ [ZYZ+18]	0.54	0.54	0.81	0.50
VALMOD [LZP+18]	0.61	0.42	0.62	0.59

on a similarity threshold that is provided by the user. However, if this information
is not available, and the threshold is set to 0.5, then the results are not as high as
the outcomes in Table 6.37

The MOEN [MuK10] method does not provide robust performance with the noisy
data, but if motifs are within the defined sliding window, this method has a higher
chance of identifying the motifs. The VALMOD [LZP+18] method cannot determine motifs that are affected by squeeze and stretch mappings.

Comparing the results of VALMOD [LZP+18] obtained from data with and without

noise reveals that this method is robust against noise. This is not the case for
MOEN Enum. [MuC15] since this algorithm discovers only one type of motifs for
some test cases such as Lightning data.

The performance of KITE variable-length motif discovery is also examined with
the noisy data. The outcomes of these investigations are given in Table 6.38.

Table 6.38: Results of KITE's variable-length motif discovery on the data covered
           with SNR of 10dB. CR: correct motif discovery rate. ED: Euclidean dis-
           tance, CD: Canberra distance, DTW: Dynamic Time Warping, EDR:
           Edit Distance on Real Sequence, LCSS: Long Common SubSequences

| | Data with Noise | | Data without Noise | |
	Best CR	Distance Measure	Best CR	Distance Measure
Gun Point	0.88	CD	0.88	ED, CD
50 Words	0.71	EDR	0.71	DTW
Food	0.83	EDR	0.98	ED
Lightning	0.62	CD	0.73	CD, EDR

In variable-length motif discovery for data without noise, EDR performs better
than other measures. In the experiments with noisy data, the CD and EDR achieve
higher results than other methods. Among the tested distance measures, EDR is
the most robust measure. Overall, the CRs of the noisy data are less in comparison
with data without noise. Appendix 11.4 provides more detail of the results obtained
from variable-length motif discovery with the synthetic data covered with noise.

An analysis of the results obtained from KITE, the MOEN Enum. [MuC15], and
VALMOD [LZP+18] algorithms regarding variable-length motif discovery is provided in Table 6.39. Both KITE and MOEN Enum. [MuC15] applied the ED
in their similarity measurement step, and VALMOD [LZP+18] employed the normalised ED. As depicted in Table 6.39, for the Gun Point, and Food data sets,
the results of KITE are superior to the MOEN Enum. [MuC15] method. However,
MOEN Enum. [MuC15] obtained better results for the 50 Words and Lightning
data sets.

Table 6.39: Variable-length motif discovery on the synthetic data sets when data
           is covered by noise (SNR=10dB).

Method/Data	Gun Point	50 Words	Food	Lightning
KITE	0.87	0.36	0.75	0.45
MOEN Enum. [MuC15]	0.58	0.45	0.52	0.52
VALMOD [LZP+18]	0.57	0.44	0.65	0.58

The MOEN Enum. [MuC15] method detects only one type of motifs in the Food,
and Lightning data sets. VALMOD [LZP+18] achieves a higher correct motif discovery rate for the Lightning data. Generally, VALMOD [LZP+18] exceeds the
MOEN Enum. [MuC15] approach and is more robust to noise.

6.6 Scalability Experiments

The scalability of KITE regarding its execution time for equal- and variable-length motif discovery is explained in this section. For the synthetic and real-world data, the execution time of KITE is illustrated in Fig. 6.16.

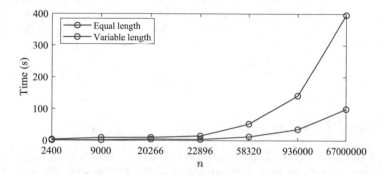

Figure 6.16: KITE's execution time for equal- and variable-length motif discovery on synthetic and real-world data.

The largest test case is AutoSense-1 with a length of $N = 67000000$, and the shortest data set is the Gun Point data set with $N = 9000$. Increasing the size of the data results in a longer execution time for KITE. The execution time, illustrated in Fig. 6.16, is gathered applying the ED since the other tested methods also employ this measure. Appendix 11.3 provides the results of the five other distance measures, which reveal that the CD achieves the minimum execution time of KITE for both equal- and variable-length motif discovery. The longest execution time is achieved when LCSS is employed.

KITE's execution time for equal-length motif discovery is also benchmarked with other state-of-the-art algorithms (cf. Appendix 11.3). Accordingly, MR.Motif [CaA10] performs faster than other algorithms, and MOEN [MuK10] is the slowest method. The MOEN Enum. [MuC15], SCRIMP++ [ZYZ+18], and VALMOD [LZP+18] algorithms are not able to detect motifs if the size of the data under investigation is large ($N > 10^6$). Thus, their execution time for AutoSense-1 is not presented.

Besides the size of the data, the motif's length affects the execution time of the MOEN [MuK10], MOEN Enum. [MuC15], SCRIMP++ [ZYZ+18], and VALMOD [LZP+18] methods. Although the size of the 50 Words data is larger than the Lightning data, the motif discovery on the Lightning data set takes more time (for all tested methods excluding Mr.Motif [CaA10]) than the 50 Words data. The reason is that the motifs of the Lightning data are longer than the 50 Words data. The execution time of KITE, when the ED is employed, is compared with MOEN Enum. [MuC15] and VALMOD [LZP+18] for variable-length motif discovery as given in Table 6.40. Correspondingly, KITE's variable-length motif discovery is more rapid than the MOEN Enum. [MuC15] and VALMOD [LZP+18] methods.

Table 6.40: The arithmetic mean value and standard deviation of the execution time of KITE, MOEN Enum., and VALMOD in variable-length motif discovery.

	KITE ($\mu \pm \sigma$)	MOEN Enum. ($\mu \pm \sigma$)	VALMOD ($\mu \pm \sigma$)
Gun Point	8.64 ± 0.05	18.31 ± 0.75	27.02 ± 0.15
Food	9.35 ± 0.07	45.81 ± 2.02	2735.25 ± 1.41
Lightning	14.31 ± 0.15	199.99 ± 15.92	1553,57 ± 0.99
50 Words	52.25 ± 0.15	491.80 ± 20.69	1721.29 ± 0.91
AutoSense-1	395.84 ± 0.23	-	-
AutoSense-2	141.29 ± 0.67	41913 ± 1.01	57400.04 ± 12.08

6.7 Case Studies

According to [Mue14], motif discovery is utilised in various applications. In this section and in Appendix 11.5, two applications of KITE are briefly described. The first application explains KITE's performance on anomaly detection, the second one provides an avenue for motif discovery in higher dimensions, given in Appendix 11.5. The work of the author published in [TDDL16, ToL17a, ToL18] is directly integrated into this section.

6.7.1 Anomaly Detection via Time Series Motif Discovery

In order to detect anomalies or classify the events of a system, one needs to learn the behaviour of that system and identify the prototype patterns for both normal and abnormal classes. Searching for the prototype patterns in a large data set can become a time and energy-consuming process for human beings. Besides, humans are not immune to making mistakes. Motif discovery facilitates this problem by detecting the prototype patterns or motifs. In order to detect attacks on ATMs and identify the normal and abnormal status of an ATM, KITE is applied in a classification task to provide motifs, which define the prototype patterns for the normal and abnormal classes.

The examples of an ATM's normal and abnormal conditions are provided in the AutoSense-1 data set (cf. Sec. 6.2.1.2). KITE's results of equal- and variable-length motif discovery for this data have been explained earlier in Sec. 6.3.2 and 6.4.2. Figure 6.17 illustrates examples of four motifs obtained by KITE during the motif discovery task. Consulting with an expert and analysing the significant motifs result in selecting motifs that appear at the beginning of the signals (motif length $l < 650000$). Thus, instead of analysing the entire signal ($N \geq 10^6$), only the signals' beginning must be considered to classify normal and abnormal events. This leads to less computational time and speeds up the classification task. Besides, this is a critical issue in on-line applications where the data is streaming [TDDL16]. After determining the motifs (prototype patterns) for each class, the classification of the non-manipulated and manipulated events is performed by the Modified-

Fuzzy-Pattern-Classifier (MFPC) [LDM04].

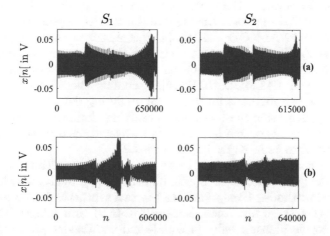

Figure 6.17: Detected motifs in non-manipulated and manipulated signals gathered
from sensors S_1 and S_2. (a) non-manipulated; (b) manipulated motifs
[TDDL16].

Table 6.41 presents the classification results of the MFPC employing 10-fold cross-
validation on the data and the Geometric mean, MFPC's original aggregation operator. These results confirm that KITE provides motifs of normal and abnormal
events, and successfully detects anomalies.

Table 6.41: MFPC's classification results of AutoSense-1 applying 10-fold cross-
validation; σ: standard deviation.

Correct classification rate (%)$\pm\sigma$	F-Measure (%)$\pm\sigma$	Sensitivity (%)$\pm\sigma$	Precision (%)$\pm\sigma$
84.0% $\pm$ 15.8	87.2% $\pm$ 15.1	83.0% $\pm$ 16.4	92.7% $\pm$ 15.7

6.8 Summary

The evaluation of KITE for different scenarios is presented in this section. The
experiments are grouped into three examinations. The first experiment's type
analyses KITE's performance for equal-length motif discovery, whereas the second
one focuses on variable-length motif discovery. Finally, the third group of experiments investigates equal- and variable-length motif discovery for noisy data. All
three groups of examinations are performed on the synthetic and real-world data
sets. Besides evaluating the outcomes by various quality measures, the results are
also benchmarked against state-of-the-art methods.

In equal-length motif discovery, KITE performed better or similar to state-of-the-art algorithms. However, if the motifs are very similar, then Mr.Motif [CaA10] outperforms the KITE method. Among state-of-the-art algorithms, only MOEN Enum. [MuC15] and VALMOD [LZP$^+$18] detect motifs of variable lengths in the same manner as KITE. Consequently, their performance is compared with KITE, where for synthetic and real-world data, KITE's performance is superior to MOEN Enum. [MuC15] and VALMOD [LZP$^+$18], excluding the 50 Words data. The structure of these two methods are based on analysing data in the time domain without any representation method and quantifying similarity by the ED. Thus, performing a representation method such as wavelet transformations and feature extraction approaches as in KITE improve the CR.

In the last investigation, in which the test cases are covered with noise, it is shown that for equal- and variable-length motifs, KITE mostly provides a higher CR than the other tested methods. Exceptions are the two synthetic data sets, 50 Words and Lightning. For these test cases, Mr.Motif [CaA10] and VALMOD [LZP$^+$18], respectively, provide higher results in equal- and variable-length motif discovery on noisy data. Comparing the results obtained from EDR in all three experiment types confirms that this method is robust to outlier and noise, and gained higher CR than other methods.

Between the investigated methods, MOEN Enum. [MuC15], VALMOD [LZP$^+$18], and MOEN [MuK10] collapse when the size of the data and the motif's length are large ($N > 10^6$). Moreover, the scalability of KITE is examined by analysing its execution time. KITE performs more quickly than other methods, even if the size of the data is increased. Nevertheless, among state-of-the-art algorithms, Mr.Motif [CaA10] is the fastest method.

Finally, this section has provided an application of KITE in classification and anomaly detection. At this point, all theoretical and practical elaborations regarding KITE have been explained. In the next section, a conclusion and an outlook of this dissertation are provided.

7 Conclusion and Outlook

After introducing the term Knowledge Discovery in Databases (KDD) by Gregory Piatetsky-Shapiro in 1989, this topic and machine learning have become an attractive area for several researchers. By increasing the power of computers to collect and compute data, the need for methods that analyse the data and extract knowledge from it is growing. This issue is addressed by tasks such as clustering, classification, query by content, anomaly detection, and motif discovery [BeR14, BLB+17, FaV17, AAJ+19, AlA20]. Motifs are previously-unknown subsequences that frequently occur in a time series or signal [PKL+02]. Motif discovery is an unsupervised and challenging task since the information about the number, length, position, or even the shape of such subsequences or motifs is not provided in advance [Mue14, ToL17b].

This dissertation is in the context of data mining and pattern recognition and is mainly focused on the discovery of motifs in time series data. Besides discussing the limitations of existing methods, a research gap is also detected and introduced in Chapter 4. Consequently, based on the state of the art, a novel method for ill-Known motIf discovery in Time sEries data (KITE) is proposed. It addresses several time series motif discovery problems such as the definition of motifs lengths, the detection of ill-known motifs, and the ability to determine motifs of both equal and variable lengths. KITE combines both procedures of pattern recognition and motif discovery, explained in Chapter 3. To the author's knowledge, KITE is the first motif discovery approach that has such a structure. KITE contains five main steps: pre-processing, representation, feature extraction, similarity measurement, and significant motif detection. KITE assigns the length of motifs either based on the expert's knowledge or automatically. If the motif's length is provided, KITE determines motifs of equal length; alternatively, it detects variable-lengths motifs. Thus, KITE solves the problem of defining the motifs' lengths in its first step.

The core of KITE is its representation step, where signals are analysed by the Analytic Complex Quad Tree Wavelet Packet Transform (ACQTWP). This wavelet transformation has properties such as shift invariance, flexible time-frequency resolution, and approximate analytic representation of a signal. Additionally, it reduces the amount of superimposed noise. Moreover, ACQTWP provides two algorithms: the early abandoning and the Best Node Selection (BNS) algorithms to select the best bases instead of analysing the complete decomposition. The early abandoning algorithm interrupts the decomposition when the shape of the input signal is not preserved by the coefficients and selects a scale with the highest amount of information content. Next, the best two nodes from the selected scale are considered by the BNS algorithm according to their energy-to-entropy ratio. This procedure reduces redundant information and accelerates the performance of ACQTWP. By

© The Author(s), under exclusive license to
Springer-Verlag GmbH, DE, part of Springer Nature 2022
S. Deppe, *Discovery of Ill-Known Motifs in Time Series Data*, Technologien
für die intelligente Automation 15, https://doi.org/10.1007/978-3-662-64215-3_7

taking advantage of the ACQTWP's characteristics, ill-known motifs altered by time-translation, stretch, and squeeze mappings, and motifs covered with noise are identified in the representation step.

The problems of ill-known motifs that are transformed by reflection mappings and have variable lengths are resolved in the feature extraction. KITE's feature extraction is flexible. This means that features can be added to or omitted from this step. Nevertheless, the maximum and minimum amount of phase and the first four statistical moments are extracted from the normalised coefficients of the ACQTWP's selected nodes. These features represent the characteristics of the signals, such as shape, variability, and central tendency, to assist motif discovery. A distance measure quantifies the similarity between the features of input signals and compares it with a threshold in order to detect motifs. If the similarity between the two subsequences is less than the provided threshold, they are considered motifs. Unlike most motif discovery approaches stated in Chapter 4, this threshold is defined automatically in KITE. Finally, misleading motifs are excluded, and K-representative motifs are determined.

The contributions of this dissertation and KITE's performance are evaluated in Chapter 6, whereby the efficiency of KITE is investigated on synthetic and real-world data. The outcomes of KITE's equal-length motif discovery are compared with the six state-of-the-art methods, and KITE's variable-length motif discovery is tested against two approaches, MOEN Enum. [MuC15] and VALMOD [LZP+18]. From the methods described in Chapter 4, only MOEN Enum. [MuC15] and VALMOD [LZP+18] identify variable-length motifs without iterating the same algorithm for several predefined motifs lengths. KITE's performance in terms of correct motif discovery rate, precision, and F-measure is superior or equal to the investigated algorithms. Additionally, KITE is robust against noise. The evaluation of the KITE's results reveals that performing a representation method such as a wavelet transformation and feature extraction as in KITE improve the results.

The outcomes of the scalability test show that extending the size of the data results in increasing the execution time for all of the tested methods. However, KITE and Mr.Motif [CaA10] complete the motif detection faster than other methods. To sum up, it is shown that KITE closes the scientific gap explained in Chapter 4.

7.1 Conclusion and Contributions

This thesis's emphasis was placed on time series motif discovery, which has been a growing research topic since its introduction in the domain of time series analysis in 2002. Various approaches have been presented to improve this field's performance and accuracy; nevertheless, as discussed in Chapter 4, besides the weak points of the existing methods, there are overlooked research issues that still raise questions. These research gaps included the definition of the length of motifs, the detection of motifs with variable lengths, huge time complexity, and the discovery of ill-known motifs.

This thesis contributes to a more stable and accurate motif discovery, resolving the dilemma of variable-lengths and ill-known motif discovery through the proposition of KITE's approach. These novel contributions made to the community of time series motif discovery are described in Chapter 5, and the most particular ones are outlined below.

1. KITE's general architecture makes it suitable for numerous motif discovery problems in various domains (cf. Chapter 6). KITE proved its performance by applying several data sets from diverse domains such as food analysis and motion capture studies.

2. KITE detects motifs of both equal and variable lengths. In contrast to the methods discussed in Chapter 4, KITE discovers variable-length motifs without iterating the same algorithm for several motif lengths. Both equal- and variable-length motifs are identified by only one iteration of KITE.

3. KITE determines ill-known motifs altered by uniform scaling, translations, stretch, and squeeze mapping. Moreover, it detects motifs that are covered with noise. This is unlike most of the explained motif discovery approaches in Chapter 4, which focus on handling one or two types of the mappings mentioned above.

4. Measuring similarities between time series signals is one of the unavoidable phases of tasks, like classification, clustering, and motif discovery. As stated in Chapters 3 and 4, applying distance measures other than the Euclidean distance is not so popular in the domain of motif discovery. In this dissertation, the performance of six different distance measures is investigated. This is the first time that the edit-based measures are employed in the motif detection task.

Besides the stated contributions, new avenues for the signal and image processing domain are explored and created:

1. Wavelet transformations are one of the most promising techniques employed in several signal and image processing applications [CLZ17,SKR17,BPC+18]. In this dissertation, the proposed wavelet transformation, ACQTWP, decomposes signals into a comprehensive time-frequency resolution, reduces the amount of superimposed noise and is approximately analytic and shift-invariant (refer to Sec. 5.3). The best bases of this transformation are selected by the early abandoning and BNS algorithms, based on the information content spread through the wavelet tree and the signal's shape. Consequently, ACQTWP is applicable in motif discovery as well as in other signal and image processing tasks.

2. Several research studies have been published addressing identifying motifs in time series signals, although determining motifs in image data sets remains in its infancy. KITE's structure is expendable for motif discovery in higher-dimensional data. In this dissertation, KITE's practicability regarding detecting image motifs is demonstrated (refer to Appendix 11.5).

To conclude, this dissertation fills the research gaps mentioned in Chapter 4 and eliminates the limitations of the existing approaches.

7.2 Perspectives and Future Directions

This section describes the research questions that arise based on the findings of this dissertation.

On-line Processing

In this thesis, off-line time series have been analysed, with a fixed length and all the data points available. However, by increasing the number of context-aware applications, the need for the real-time decision making and the analysis based on the information derived from sensors also increases [DTC$^+$18, RJV18, JZP$^+$19, CRF$^+$19]. Therefore, on-line algorithms that extract information from streaming data are required. In such methods, the data is segmented by a sliding window, where the most recent data are in the current window [HCZ$^+$13, Spi15, JZP$^+$19]. On-line motif discovery approaches are not so eminent due to certain challenges. This includes the development of methods that operate in a considerable amount of time, handle multi-dimensional data obtained from multiple sensors, detect motifs not only of equal but also variable lengths, and provide the length of motifs even without expert knowledge. Few motif discovery methods exist that can handle the problem of streaming data in on-line applications. Methods proposed in [LiL10, MuK10, CaA10, JZP$^+$19, LiL19] can detect motifs in on-line applications. These methods employ SAX [LKL$^+$03] and its versions in their representation step which as stated and explained in [BuK15] has several problems. Recent approaches [ZYZ$^+$18, MIM$^+$19, ImK19, ZYZ$^+$20] that are based on the concept of matrix profile also are able to detect motifs in streaming time series. However, in order to detect similarity, the Euclidean distance, which its drawbacks are mentioned, is employed in these methods. Additionally, the majority of existing on-line motif discovery methods can determine motifs of equal length.

Thus, novel approaches are needed to detect variable-length and ill-known motifs in on-line applications. KITE is executable for the equal-length motif discovery in on-line applications due to the concept of sliding windows employed in its pre-processing step. Consequently, the problem of identifying ill-known motifs for on-line motif discovery can be handled by KITE. Nevertheless, providing an on-line variable-length motif discovery is intended as one area of future work.

Multivariate Motif discovery

Multivariate time series motif discovery has gained attention during the last decade. Multivariate time series are part of the domains like biomedical (e.g. EEG signals), geophysical (e.g. monitoring earthquakes), and sensor network. The first proposed method employed HMM [BaP66] in its representation step to model the motifs' statistics [Oat02]. Other methods, as in [MIE$^+$07a, MIE$^+$07b, VAS09, Bev12], apply

PCA [Pea01] to project the multivariate signal into the univariate signal and then use random projection [BuT01] and SAX [LKL$^+$03] to detect motifs. Recently, the matrix profile based methods as [ZYZ$^+$18] are able to detect motifs in multivariate time series by MDL [Gru05].

Two main shortcoming of these techniques are the motif length, which must be provided by the user, and identifying motifs in the same time stamp. Moreover, most of the described methods have quadratic computational complexity.

Research on variable-length motif discovery for multivariate time series is limited. In [BWP16], a grammar based framework [LLO12] is employed to detect variable-length motifs in healthcare multivariate signals. In this approach motifs are detected in each dimension and after that motifs in different dimensions are grouped based on their timestamps. According to [GaL19], this method is mostly suitable for low-dimensional time series. Recently, Gao and Lin [GaL19] presented a method for variable-length multivariate motif discovery which is based on their previous approach [GaL18] and SAX [LKL$^+$03]. Although their method identifies variable-length motifs in multivariate signals, several parameters (for SAX) must be defined in advance. This approach has a quadratic time complexity ($\mathcal{O}(N^2)$, where $N \in \mathbb{N}$ is the length of the time series). Despite the time cost of this method, the authors claim that their algorithm handles million size multivariate time series [GaL19]. The stated approaches aim to find motifs in multivariate signals that are located in the same time range. Further research must be performed to investigate if determining motifs in different time slots in multivariate data provides meaningful information. The most important drawback of these methods is their inability to detect ill-known motifs.

This problem can be tackled by KITE, which can be employed to discover variable-length motifs in multivariate signals. Both representation and feature extraction steps of KITE are able to handle multi-dimensional data, as explained in Appendix 11.5. Nevertheless, it should be analysed which procedure provide more accurate results: detecting variable-length motifs in each dimension separately or considering time series in all dimensions at once.

Image Motif Discovery

During the last decade, the boost of optical imaging technologies results in collecting more data with a faster rate. Most of these data are in the form of graphics, pictures, videos or integrated multimedia [ToL18]. Similar to time series analysis, machine learning and data mining tasks analyse and provide efficient information from such data. Tasks such as clustering or classifying images as well as finding the query images in an image database have been investigated during last decades [Tya17, LAF$^+$18, OZH$^+$10]. However, the problem of deriving structures or detecting regularities in image databases is not fully investigated by researches [ToL18]. Such frequently occurring structures in an image or similar images within an image database, which are unknown due to lack of prior information, are called *image motifs*.

Detecting image motifs is a growing research topic with diverse applications such as identifying motifs in historical documents and human social behaviour analysis. Most of the proposed methods for image motif discovery convert images into one dimensional time series by approaches such as contouring [XKW+07] and then attempt to de motifs in such data [YeK09,CFC+12,GSST16,VDV20]. Transforming a two dimensional data to a one dimensional usually lead to information loss, which is the main drawback of these methods. Additionally, identifying ill-known motifs is still a challenging problem for these approaches. Methods such as [RZK11,EPN+15] analyse images in their original form by segmenting images using a sliding window of a fixed size. This procedure leads to inability of determining motifs with various proportions [ToL18].

KITE's approach can be extended to detect image motifs. Preliminary research has been conducted and published by the author in [ToL18]. KITE's image motif discovery investigates images in databases of images and shapes without transforming the images to one-dimensional time series. The same structure for one dimensional data is applied to determine image motifs with KITE, nevertheless, the extension of the ACQTWP transform, the 2D-ACQTWP transform, is employed, as explained in Appendix 11.5. The obtained results of KITE for image motif discovery in image databases showed that KITE overcomes the drawbacks of the aforementioned methods [ToL17a,ToL18]. However, KITE cannot determine all types of mappings and distortions. As an example, motifs altered by rotation mappings in higher-dimensional data cannot be identified by KITE since 2D-ACQTWP is not rotation invariant. Thus, providing a rotation-invariant property for the 2D-ACQTWP transformation is considered as future work.

Further research must be preformed in order to analyse the performance of KITE for image motif discovery within an image. An extension of KITE will be to locate motifs within an image without segmenting the tested image into various sections by, e.g. employing the new approach, given in [RaH20], to measure image similarity.

8 Appendix A

8.1 Function and Signal Space

A linear vector space whose vectors are functions is called a *function space*. In this space, scalers are real numbers, and the inner product is obtained by [BGG⁺98, PoM07]

$$b = \langle f(t), g(t) \rangle = \int f^*(t)g(t)dt, \tag{8.1}$$

where b is a scalar, and $f(t)$ and $g(t)$ are two functions. The norm of a function $f(t)$ is given by $\|f\| = \sqrt{|\langle f, f \rangle|}$ [PoM07]. The Hilbert space $L^2(\mathbb{R})$ in signal processing consists of all functions $f(t)$ with a finite and well defined integral of the square: $f \in L^2 \Rightarrow \int |f(t)|^2 dt = E < \infty$ [BGG⁺98, PoM07].

Definition 8.1 (Expansion set). "If any function $f(t)$ in the vector space of signals $\mathcal{S}$, $f(t) \in \mathcal{S}$, is represented by $f(t) = \sum_k a_k \psi_k(t)$, then the set $\psi_k(t)$ is an expansion set for $\mathcal{S}$" [Fri41, BGG⁺98].

The expansion set is a basis if the series representation is unique [BGG⁺98].

Definition 8.2 (Span of a basis set). "If the space $\mathcal{S}$ can be defined by the set of all functions expressed by $f(t) = \sum_k a_k \psi_k(t)$, where $\psi_k(t)$ is the expansion or basis set, then this is called the span of the basis set" [BGG⁺98, PoM07].

8.2 Transformations and Representation

Time series can be analysed in their original domain or investigated in another representation or transform domain. Transformations usually help to approximate the data without losing information.

Definition 8.3 (Transformation). A bijective function or mapping $T : A \to B$ is a transformation if it is both surjective and injective [Woo96].

Definition 8.4 (Invariant transformation). A transformation T is called invariant to the group action $\mathcal{G}$ if [BuS95]:

$$\forall g \in \mathcal{G}, \ x \in X \ \cdot T(gx) - T(x).$$

Definition 8.5 (Affine transformation). An affine transformation $T_\mathrm{A} : A \to B$ is a mapping of the form $T_\mathrm{A}(a) = aM + b$, where M is a linear transformation on the space A, and $a \in A$, $b \in B$ [Ber09, BuB09].

© The Editor(s) (if applicable) and The Author(s), under exclusive license to
Springer-Verlag GmbH, DE, part of Springer Nature 2022
S. Deppe, *Discovery of Ill-Known Motifs in Time Series Data*, Technologien
für die intelligente Automation 15, https://doi.org/10.1007/978-3-662-64215-3

Translation, scaling, and reflection are examples of affine transformations.

Definition 8.6 (*Time series representation*). The time series representation or mapping $T : \mathbb{R} \to \mathbb{R}$ transforms a given time series $x[n]$ with $1 \le n \le N$, $N \in \mathbb{N}$ to $\hat{x}[n]$, where $\hat{x}[n]$ closely approximates $x[n]$ [EsA12].

One important aspect of time series representations or mappings is the invariant property.

Definition 8.7 (Invariant representation). An invariant representation or transformation T maps all time series of an equivalence class under the transformation group $\mathcal{G}$ into one point in the transformation space. Thus, for two time series x_1 and x_2, the following holds [BuS95]

$$x_1, x_2 \in X, \; g \in \mathcal{G}, \; T(gx_1) \cong T(gx_2).$$

Thus, two signals x_1 and x_2 are equivalent under the group $\mathcal{G}$ if $x_1 \cong x_2 \; \Rightarrow \; T(x_1) \cong T(x_2)$.

8.3 Wavelet Transform

Wavelet [Dau90, Mey93, BGG$^+$98, JeC01, LLZ$^+$02, Mal08, Pou10], $\psi(t)$, refers to a function with zero average, and the energy concentrated in time $\int_{-\infty}^{\infty} \psi(t)dt = 0$. Signal $x(t)$ can be analysed by a set of expansion functions $\psi_{j,k}(t)$ such as [Mey93]

$$x(t) = \sum_{j,k} a_{j,k} \psi_{j,k}(t), \tag{8.2}$$

where $a_{j,k}$ is a set of coefficients (wavelet coefficients), and parameters $j, k \in \mathbb{Z}$ are scaling and translation parameters of the mother wavelet $\psi(t)$. Eq. 8.2 can be rewritten as [Mey93, BGG$^+$98]

$$x(t) = \sum_{j,k} a_{j,k} 2^{j/2} \psi(2^j t - k). \tag{8.3}$$

In order to describe the wavelet transformations, the concepts of resolution and a scaling function $\phi(t)$ are required.

8.3.1 Scaling Function

A set of scaling functions is given by [BGG$^+$98]

$$\phi_k(t) = \phi(t - k), \qquad k \in \mathbb{Z}, \qquad \phi \in L^2,$$

where $\phi(t - k)$ includes time-translated versions of the basic scaling function. Let V_0 be the space, called the approximation space, spanned by these functions, then any signal $x(t)$ can be defined by $x(t) = \sum_k a_k \phi_k(t)$. Assigning different values to the scaling and translation parameters results in a family of scaling functions, such that $\phi_{j,k}(t) = 2^{j/2}\phi(2^j t - k)$, with the span over $k \in \mathbb{Z}$, $\mathcal{V}_j = \overline{\mathrm{Span}\{\phi_k(2^j t)\}} = \overline{\mathrm{Span}\{\phi_{j,k}(t)\}}$ [BGG$^+$98, Ada10].

8.3.2 Multiresolution Analysis

A multiresolution analysis (MRA) consists of a nesting of successive approximation spaces $\mathcal{V}_j$, $j \in \mathbb{Z}$, with the four properties given below [Mal89, BGG$^+$98, Ada10]

- $\cdots \subset \mathcal{V}_{-1} \subset \mathcal{V}_0 \subset \mathcal{V}_1 \subset \mathcal{V}_2 \subset \cdots \subset L^2$,

- $\mathcal{V}_j \subset \mathcal{V}_{j+1}, \forall j \in \mathbb{Z}$,

- $\mathcal{V}_{-\infty} = 0, \qquad \mathcal{V}_\infty = L^2$,

- $x(t) \in \mathcal{V}_j \iff x(2t) \in \mathcal{V}_{j+1}$.

The first three conditions warranty that the space contains high-resolution signals, also includes lower resolution signals. The fourth constraint ensures that the elements in the current space are scaled versions of the elements in the next space [Mal89, BGG$^+$98]. A graphical representation of the nested spaces is illustrated in Fig. 8.1.

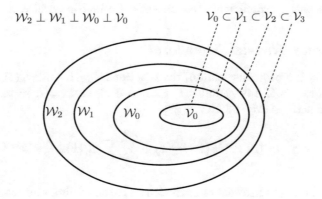

Figure 8.1: Nested spaces spanned by the expansion set of scaling functions [BGG$^+$98]

For the given multiresolution approximation space $\mathcal{V}_j$, there exists a scaling function $\phi(t)$, which is an orthonormal basis of $\mathcal{V}_j$. Based on the above constraints, if $\phi(t) \in \mathcal{V}_0$ then the nesting of the spans $\phi(2^j t - k)$ is given by $\mathcal{V}_j$. Consequently, if $\phi(t) \in \mathcal{V}_0$, then it is a member of $\mathcal{V}_1$, the space spanned by $\phi(2t)$, and so [BGG$^+$98, Pou10]

$$\phi(t) = \sum_n g(n)\sqrt{2}\phi(2t - n), \qquad n \in \mathbb{Z}, \tag{8.4}$$

where $g(n)$ denotes some set of coefficients (coefficients of a low-pass filter) and to obtain the norm of the scaling function with the scale of two, factor $\sqrt{2}$ is applied [BGG$^+$98].

The detail signal at scale 2^j is obtained from the difference between the approximation of a signal $x(t)$ at scales 2^{j+1} and 2^j. This signal can be described by a set of functions $\psi_{j,k}(t)$ that span the differences between approximation spaces, which are built by various scales of the scaling function [BGG$^+$98, Ada10].

If the scaling and wavelet functions are orthogonal, then their advantages are a straightforward calculation of the expansion coefficient, and profiting from the Parseval's Theorem (cf. Theorem 3). The orthogonal complement of $\mathcal{V}_j$ in $\mathcal{V}_{j+1}$ is denoted by $\mathcal{W}_j$, cf. Fig. 8.1. Thus, $\mathcal{W}_j$ is orthogonal to $\mathcal{V}_j$ means that $\mathcal{V}_{j+1} = \mathcal{V}_j \oplus \mathcal{W}_j$, and [BGG$^+$98, Mal89]

$$L^2 = \mathcal{V}_0 \oplus \mathcal{W}_0 \oplus \mathcal{W}_1 \oplus \cdots, \qquad (8.5)$$

where $\mathcal{V}_0$ is the initial space spanned by the scaling function $\phi(t - k)$. To sum up, Wavelet members of $\mathcal{W}_0$ are located in the space of the next narrower scaling function $\mathcal{V}_0$ [BGG$^+$98] (cf. Fig. 8.1). Therefore, they can be obtained from a weighted sum of shifted scaling function $\phi(2t)$, such that [BGG$^+$98, Mal08]

$$\psi(t) = \sum_n h(n)\sqrt{2}\phi(2t - n), \qquad n \in \mathbb{Z}, \qquad (8.6)$$

where $h(n)$ is a set of coefficients (coefficients of a high-pass filter).

8.3.3 Discrete Wavelet Transform

As explained in the previous section, the functions $\psi_{j,k}(t)$ and $\phi_k(t)$ span all of L^2 and based on Eq. 8.5, any signal $x(t) \in L^2$ can be expressed in terms of scaling and wavelet functions [Mey93, BGG$^+$98]

$$x(t) = \sum_k c_{j_0}(k)2^{j_0/2}\phi(2^{j_0}t - k) + \sum_{j=0}^{\infty}\sum_k d_j(k)2^{j/2}\psi(2^j t - k), \qquad (8.7)$$

where $c_{j_0}(k)$ and $d_j(k)$ are sets of coefficients. If the scaling and wavelet functions are orthogonal, these coefficients can be calculated using inner products [Mey93, BGG$^+$98]

$$c_j(k) = \langle x(t), \phi_{j,k}(t) \rangle = \int x(t)\ \phi_{j,k}(t)\ dt,$$
$$d_j(k) = \langle x(t), \psi_{j,k}(t) \rangle = \int x(t)\ \psi_{j,k}(t)\ dt. \qquad (8.8)$$

In Eq. 8.7, the first summation term provides a function that is a low-resolution approximation of the signal $x(t)$. The second summation term denotes a higher-resolution function, which adds details as index j grows. Thus, $c_j(k)$ provides the scaling or approximation coefficients, and the wavelet or detail coefficients are obtained by $d_j(k)$ [BGG$^+$98].

Eq. 8.7 describes the discrete wavelet transform (DWT) [Mey93].

8.3.4 Filter Bank Structure

Instead of applying the scaling functions and wavelets in order to analyse a signal, it is possible to employ the filter bank structure [Mal08]. This structure was first introduced by Mallat [Mal08] and is also called Mallat's implementation, which considers the low and high-pass filters, $g(n)$ and $h(n)$, and the scaling and wavelet coefficients $c(k)$ and $d_j(k)$.

The concept of the filter bank structure is based on the relationship between the expansion coefficients at a lower scale in terms of those at a higher scale [Mal08]. According to Eq. 8.4

$$\phi(t) = \sum_n g(n)\sqrt{2}\phi(2t - n), \tag{8.9}$$

where scaling and translating the variable t results in [BGG$^+$98, Mal08]

$$\phi(2^j t - k) = \sum_n g(n)\sqrt{2}\phi(2(2^j t - k) - n) = \sum_n g(n)\sqrt{2}\phi(2^{j+1}t - 2k - n).$$

Setting $m = 2k + n$ leads to $\phi(2^j t - k) = \sum_m g(m - 2k)\sqrt{2}\phi(2^{j+1}t - m)$. Now, let $\mathcal{V}_j = \text{Span}\{2^{j/2}\phi(2^j t - k)\}$ then $x(t) \in \mathcal{V}_{j+1} \Rightarrow x(t) = \sum_k c_{j+1}(k)2^{(j+1)/2}\phi(2^{j+1}t - k)$. Thus, at scale $j + 1$, $x(t)$ is expressible with scaling functions, however at one scale lower j, wavelet functions are required for the detail information, such as [Mal08, Ada10]

$$x(t) = \sum_k c_j(k)2^{j/2}\phi(2^j t - k) + \sum_k d_j(k)2^{j/2}\psi(2^j t - k). \tag{8.10}$$

As already stated, if the wavelet and scaling functions are orthogonal, then the scaling and wavelet coefficients $c_j(k)$ and $d_j(k)$ are obtained by the inner product [BGG$^+$98].

$$c_j(k) = \langle x(t), \phi_{j,k}(t) \rangle = \int x(t)\, 2^{j/2}\, \phi(2^j t - k)\, dt$$

$$= \sum_m g(m - 2k) \int x(t)\, 2^{j+1/2}\, \phi(2^{j+1}t - m)\, dt$$

and so [Mal08]

$$c_j(k) = \sum_m g(m - 2k)c_{j+1}(m). \tag{8.11}$$

Similarly, the wavelet coefficients can be calculated [Mal08]

$$d_j(k) = \sum_m h(m - 2k)c_{j+1}(m), \tag{8.12}$$

where $g(n)$ and $h(n)$ are low and high-pass filters, respectively. Based on Eq. 8.8

$$c_j(k) = \langle x(t), \phi_{j,k}(t) \rangle,$$
$$d_j(k) = \langle x(t), \psi_{j,k}(t) \rangle. \qquad (8.13)$$

The above equations Eq. 8.11 and 8.12 can be represented by a decomposition scheme (2-band filter bank) as in Fig. 8.2. Consequently, this 2-band structure applies low and high-pass filters in each scale and a downsampler. Downsamplers take a signal $x(n)$ as an input and provide an output $y(n) = x(kn)$, where $k \in \mathbb{Z}$ is called a downsampler factor. This factor is mostly equal to two (as in this work), but it can take any other value as well.

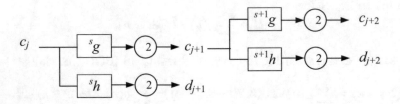

Figure 8.2: Two scales DWT decomposition scheme applying the filter bank structure [BGG+98].

The filter banks' first scale divides the spectrum of $c_{j+1}(k)$ into two equal parts of a low-pass and high-pass band. In the second scale, the low-pass band is partitioned into another lower low-pass and high-pass band. So in this scale, the lower half band of the previous scale is divided into quarters. This procedure continues for the rest of the scales resulting in the frequency bands depicted in Fig. 8.3, where the frequency response of a filter $g(n)$ or $h(n)$ is given by the discrete-time Fourier transform of its impulse response $H(\omega)$.

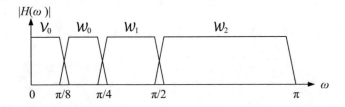

Figure 8.3: Frequency bands for the 3 scales two-bands filter bank tree [BGG+98].

The explained structure in this section represents an *analysis* filter bank [Mal08]. It is possible to reconstruct the signal by the *synthesis* filter bank [Mal08] or the inverse decomposition.

8.3.5 Synthesis Filter Bank

For synthesis filter bank, the approximation and detail coefficients are combined to reconstruct the original signal. For DWT, this transformation is called the inverse discrete wavelet transform (IDWT) [Mey93, BGG+98, Mal89].

If signal $x(t)$ is in the $j+1$ scaling function space $x(t) \in V_{j+1}$, then it can be written in terms of the scaling function as [BGG+98]

$$x(t) = \sum_k c_{j+1} 2^{(j+1)/2} \phi(2^{j+1}t - k). \tag{8.14}$$

In the next scale, $x(t)$ can be calculated by [BGG+98]

$$x(t) = \sum_k c_j 2^{j/2} \phi(2^j t - k) + \sum_k d_j(k) 2^{j/2} \psi(2^j t - k).$$

Replacing Eq. 8.9 in the above equation leads to

$$x(t) = \sum_k c_j \sum_n g(n) 2^{(j+1)/2} \phi(2^{j+1}t - 2k - n) + \sum_k d_j(k) \sum_n h(n) 2^{(j+1)/2} \phi(2^{j+1}t - 2k - n). \tag{8.15}$$

Finally, multiplying Eq. 8.14 and 8.15 by $\phi(2^{j+1}t - k)$ and integrating them results in the inverse transformation [BGG+98, Pou10]

$$c_{j+1}(k) = \sum_m c_j(m) g(k - 2m) + \sum_m d_j(m) h(k - 2m). \tag{8.16}$$

It should be noted that the above multiplication and integration are obtained if all the above wavelet and scaling functions are orthonormal.

9 Appendix B

9.1 Nobel Identities

Nobel identity shows the way that the position of down/upsamplers and filters can be interchanged [Vai90], as illustrated in Fig. 9.1.

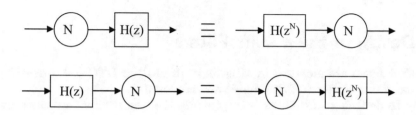

Figure 9.1: Noble Identity [Vai90], interchanging the positions of down/up sampling filters.

Assume that signal $y[n]$ is the output of the convolution of the impulse response of filter $h[n]$ and the signal $x_1[n]$. The output signal $y[n]$ is obtained by

$$y[n] = x_1[n] * h[n]$$
$$= \sum_{k=-\infty}^{\infty} x_1[n]h[n-k]. \tag{9.1}$$

Eq. 9.1 can be written as

$$= \sum_{l=-\infty}^{\infty} h[l]x_1[n-l]. \tag{9.2}$$

Replacing $x[2n]$ instead of $x_1[n]$ in Eq. 9.2 results in [Sut97]

$$= \sum_{l=-\infty}^{\infty} h[l]x[2n-2l]. \tag{9.3}$$

Here the index $2n$ is due to downsampling. Replacing $2n$ by n indicates the equivalent system at the point after the filter and results in $\sum_{l=-\infty}^{\infty} h[l]x[n-2l]$. This means that $h[l]$ is zero at any places other than $2l$th point [Vai90]

$$h_1[n] = \begin{cases} 0 & n \text{ is odd,} \\ h[\frac{n}{2}] & \text{otherweise.} \end{cases}$$

Hence,

$$y[n] = \sum_{l=-\infty}^{\infty} h_1[l]x[2n-l], \qquad (9.4)$$

which demonstrates the impulse response of an equivalent filter before downsampler is the impulse response of the original filter but upsampled by two, as depicted in Fig. 9.1 [Vai90].

9.2 Design of the q-shift Filters

The q-shift filters are proposed by Kingsbury [Kin01] for DTCWT, where Conjugate Quadrature Filters (CQF) [Mal08] are employed.
The key to designing such filters is to generate the tree B filters are time reverse of the tree A filters [Kin01]. Thus,

$$G_b(z) = z^{-1}G_a(z^{-1}), \qquad (9.5)$$

where $G(z)$ is a FIR filter of order $M = 2n$, $n \in \mathbb{N}$ and its transfer function is given by $G(z) = \sum_{n=0}^{M} g(n)z^{-n}$ [Zha11]. The filters have real coefficients, thus Eq. 9.5 satisfies

$$|G_b(e^{j\omega})| = |G_a(e^{j\omega})|, \qquad \angle G_b(e^{j\omega}) = -M\omega - \angle G_a(e^{j\omega}).$$

The necessary half-sample delay condition (cf. Theorem 1) is achieved by [Kin03]

$$\begin{aligned} \angle G_a(e^{j\omega}) &\simeq -\frac{M\omega}{2} + \frac{\omega}{4}, \\ \angle G_b(e^{j\omega}) &\simeq -\frac{M\omega}{2} - \frac{\omega}{4}. \end{aligned} \qquad (9.6)$$

As explained in [Kin03], filters $g_a[n]$ and $g_b[n]$ must have a group delay of $1/4$ sample $(+q)$ relative to the midpoint of the filters at $M/2$ sample periods. Hence, tree A filters are time reverse of the tree B filters so that the filters in tree B have a delay of $3q$.
In order to design the low-pass filter $G(z)$ of length $M = 2n$ with the aforementioned properties, Kingsbury proposed to design a low-pass filter $G_{L2}(z)$ of length $2M = 4n$, so that [Kin01]

$$G_{L2}(z) = G_a(z^2) + z^{-1}G_b(z^{-2}). \qquad (9.7)$$

The delay property of $G_{L2}(z)$ is illustrated in Fig. 9.2, in which the filter samples of the two filters G_a and G_b are interleaved to form a single smooth low-pass filter of even length and with symmetric filter coefficients [Che09].

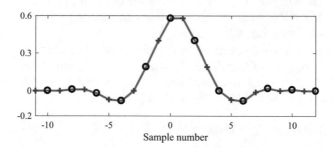

Figure 9.2: Impulse response of the oversampled $G_{L2}(z)$ for $n = 6$. The samples of G_a and G_b are shown as circles and crosses. Both G_a and G_b filters consist of $2n = 12$ taps [Kin03].

Besides the half-sample delay condition, filters G_a and G_b must satisfy the perfect reconstruction condition given by [BGG$^+$98, Kin01]

$$G_a(z)G_a(z^{-1}) + G_a(-z^{-1})G_a(-z) = 2$$

Note: The relation between high-pass and low-pass filters of type CQF is given by $H(z) = z^{-(M-1)}G(-z^{-1})$.

Additionally, q-shift filters should be orthogonal to provide a tight frame transform [BGG$^+$98]. This has an advantage of conserving energy from the signal in the transform domain (Parseval's theorem). Furthermore, the same filters and their time-reversed can be used in both tree A and B, and in the analysis and synthesis wavelet transforms.

Several methods are introduced in the literature for designing the q-shift filters [TKP06, Zha11]. The two main approaches are explained in the following [Che09]

1. Zero-forcing methods:

These approaches employ the Bernstein polynomial [CaA93]

$$B_N(x, \alpha) = \sum_{i=0}^{(N-1)/2} (1-\alpha_i)\binom{N}{i}x_i(1-x)^{(N-i)} + \sum_{i=(N+1)/2}^{N} \alpha N - i\binom{N}{i}x_i(1-x)^{(N-i)},$$

where $\alpha = (\alpha_0, \alpha_1, ..., \alpha_{(N+1)/2})^T$. If $N = 2K + 1$, then only one degree of freedom (α_K) must be adjusted in order to optimise and design the q-shift filter delays [Che09]. In [TKP06], the authors evaluated CQFs of length 4 to 22. Note that all the filters are of even length since the filter length can only be even in order to have orthogonality [Mal08].

The following values of α_K, 0.0460, 0.1827, 0.2405, and 0.0245, result in filters of lengths 8, 12, 18, and 22, respectively. The authors suggested applying filters of longer length (greater than 8) since they are smoother than filters of length less or equal to 8. Refer to [TKP06] for further information.

2. Iterative energy minimization methods:

This method, given in [Kin01], provides the filter $G_a(z)$ and its time-reversed $G_b(z)$ so that they satisfy the standard CQF conditions [Mal08] and also give a smooth low-pass filter applying Eq. 9.7 [Che09].

Based on the perfect reconstruction and orthogonality conditions [Dau90, Mal89, BGG+98] $G_a(z)H_a(z) = G_a(z)G_a(z^{-1})$ must have no terms in z^{2k} except in z^0. Thus, $G_a(z^2)G_a(z^{-2})$ must have no terms in z^{4k} except in z^0. According to Eq. 9.7, $G_{L2}(Z) = G_a(z^2) + z^{-1}G_a(z^{-2})$, and so [Che09]

$$G_{L2}(z)G_{L2}(z^{-1}) = 2G_a(z^2)G_a(z^{-2}) + z^{-1}G_a^2(z^{-2}) + zG_a(z^2).$$

Thus, $G_{L2}(z)G_{L2}(z^{-1})$ must also have no terms in z^{4k} except in z^0. Now, one needs to design the low-pass filter $G_{L2}(z)$ of length $4n$ such that [Kin03, Che09]:

- Zero amplitude for all the terms of $G_{L2}(z)G_{L2}(z^{-1})$ in z^{4k} except in z^0. This provides quadratic constraints for the filter coefficients of g_{L2}.

- Zero (or near-zero) amplitude of $G_{L2}(e^{j\omega})$ for the stopband, $\pi/3 \le \omega \le \pi$. This sets linear constraints on G_{L2}.

The above conditions result in a set of equations for $2n$ unknowns that form one half of the symmetric filter g_{L2} (the length of filters g_a and g_b is equal to $2n$).

By applying the matrix pseudo-inverse method, the least mean square (LMS) error solution can be detected. Nevertheless, this requires that all the constraints are linear [Che09]. The quadratic constraints are linearised considering an iterative solution, so that G_{L2} at iteration $i \in \mathbb{N}$ is given by $g_i = g_{i-1} + \Delta g_i$, then [Kin03]

$$\begin{aligned} g_i * g_i &= (g_{i-1} + \Delta g_i) * (g_{i-1} + \Delta g_i) \\ &= g_{i-1} * (g_{i-1} + 2\Delta g_i) + \Delta g_i * \Delta g_i. \end{aligned} \tag{9.8}$$

As i increases, Δg_i becomes small, thus the final term in above equation can be neglected and the convolution $(*)$ is demonstrated as a linear function of Δg_i. The problem of designing g_a is expressed in the following by solving Δg_i so that [Che09]

$$C(g_{i-1} + 2\Delta g_i) = (0, ..., 0, 1)^T, \tag{9.9}$$

$$F(g_{i-1} + \Delta g_i) \simeq (0, ..., 0)^T, \tag{9.10}$$

where matrix C computes every 4th term of convolution with g_{i-1} and matrix F calculates the Fourier transform at $8n$ discrete frequencies $\pi/3 \le \omega \le \pi$. $8n$ ensures that all sidelobe maxima and minima are captured reasonably accurately [Che09]. As stated in [Kin03], only one side of the convolution is needed in C, since the result is symmetric about the central term. Moreover, the columns of matrices C and F can be combined in pairs so that it is enough to solve only the first half of the symmetric Δg_i [Che09].

To obtain high accuracy solutions, it is possible for an iterative least mean square framework to scale up by a factor $\beta_i = 2^i$. Thus, Eq. 9.9 is expressed by [Kin01]

$$\begin{bmatrix} 2\beta_i C \\ F \end{bmatrix} \Delta g_i = \begin{bmatrix} \beta_i(c - Cg_{i-1}) \\ -Fg_{i-1} \end{bmatrix} \qquad (9.11)$$

$$g_i = g_{i-1} + \Delta g_i, \qquad (9.12)$$

where $c = (0, ..., 0, 1)^T$. In [Kin01], by setting $i = 20$, filters of lengths 10, 14 and 18 are obtained. Table provides the coefficients of g_a for $n = 5, 7$ and 9.

Table 9.1: Filter coefficients of $G_a(z)$ for $n = 5, 7, 9$ [Kin01].

6-tap g_a	14-tap g_a	18-tap g_a
		-0.0022
		0.0012
	0.0032	-0.0118
	-0.0038	0.0012
0.0351	0.0346	0.0443
0	-0.0388	-0.0532
-0.0883	-0.1172	-0.1133
0.2338	0.2752	0.2809
0.7602	0.7561	0.7528
0.5875	0.5688	0.5658
0	0.0118	0.0245
-0.1143	-0.1067	-0.1201
0	0.0238	0.0181
0	0.0170	0.0315
	-0.0054	-0.0066
	-0.0045	-0.0025
		0.0012
		0.0024

The Matlab M-files for designing q-shifts is provided by Kingsbury in `http://sigproc.eng.cam.ac.uk/Main/NGK`.
Kingsbury q-shift filters have two vanishing moments. Zhang [Zha11] proposed a method to design q-shift filters with improved vanishing moments. In this thesis, q-shift filters of length 14 taps are applied due to the comparison performed on several q-shift filters in [Kin01].

9.3 Proof Lemma 1

The first scale wavelet coefficients of signal $x[n - S_o]$ for odd shifts are given in following:
Odd Shifts. If $x[n - S_o]$ where the shift $S_o = 2m + 1$ then the real wavelet

coefficients are computed

$$\text{Re}(\,^1d_o[n, S_o]) = \sum_{k=0}^{L+M-1} x[2n - k - S_o]\,^1h_a[k] = \cdots$$

$$\cdots = \sum_{k=0}^{L+M-1} x[2n - k - 2m - 1 + (1-1)]\,^1h_a[k] \stackrel{^1h_a[k]=\,^1h_b[k-1],\ p=k-1}{=\!=} \cdots$$

$$\cdots = \sum_{p=-1}^{L+M-2} x[2(n - m - 1) - p]\,^1h_b[p],$$

since the high-pass filter $^1h_b[p] = 0$ for $p < 0$ and also based on the convolution property ($\,^1d_o[n, S_o]) = 0$ when $p > M + L - 1$, then

$$\cdots = \sum_{p=0}^{L+M-1} x[2(n - m - 1) - p]\,^1h_b[p] = \text{Im}(\,^1d[n - m - 1])$$

The imaginary coefficients are obtained by

$$\text{Im}(\,^1d_o[n, S_o]) = \sum_{k=0}^{L+M-1} x[2n - k - S_o]\,^1h_b[k] = \cdots$$

$$\cdots = \sum_{k=0}^{L+M-1} x[2(n - m) - k - 1]\,^1h_b[k] \stackrel{p=k+1,\ ^1h_a[k]=\,^1h_b[k-1]}{=\!=} \cdots$$

$$\cdots = \sum_{p=0}^{L+M-1} x[2(n - m) - p]\,^1h_a[p] = \text{Re}(\,^1d[n - m]).$$

From above equations, it can be concluded that,

$$\text{Re}(\,^1d_o[n, S_o]) = \text{Im}(\,^1d[n - m + 1]) = \text{Im}(\,^1d[n - (\lfloor \tfrac{S_o}{2} \rfloor) + 1]),$$

$$\text{Im}(\,^1d_o[n, S_o]) = \text{Re}(\,^1d[n - m]) = \text{Re}(\,^1d[n - (\lfloor \tfrac{S_o}{2} \rfloor)]).$$

Even Shifts. The same procedure proves the results for even shifts.

9.4 Proof Lemma 2

Even Shifts. Let $s > 1$, and $S_e = 2m$, where $m = \{m | \exists \beta,\ m = 2^{s-1}\beta + \alpha,\ \beta \in \mathbb{Z}, 0 \leq \alpha \leq s\}$ then for

- $\alpha = 0$, $m = 2^{s-1}\beta$, thus

$$\text{Re}(\ ^s d_e[n, S_e]) = \sum_{k=0}^{L+M-1} x[2^s n - k - 2^I \beta]\ ^2 h_a[k] = ...$$

$$... = \sum_{k=0}^{L+M-1} x[2^s(n - \beta) - k]\ ^2 h_a[k] = \text{Re}(\ ^2 d[n - \beta]).$$

- $\alpha \geq 1$, $m = 2^{s-1}\beta + \alpha$. Consequently

$$\text{Re}(\ ^s d_e[n, S_e]) = \sum_{k=0}^{L+M-1} x[2^s n - k - 2^s \beta - 2s]\ ^2 h_a[k] = ...$$

$$... = \sum_{k=0}^{L+M-1} x[2^s(n - \beta) - k - 2s]\ ^2 h_a[k] \neq \text{Re}(\ ^s d[(n)]).$$

Therefore, $\text{Re}(\ ^s d[(n)]) \ncong \text{Re}(\ ^s d_e[n, S_e])$. Imaginary coefficients provide the same outcome.

Odd Shifts. The same procedure proves the results of odd shifts.

10 Appendix C

10.1 Proof of Parseval's Theorem

Proof. Recall the multiresolution property and the orthogonal complement W_j, then W_j is orthogonal to V_j and $W_j \oplus V_j = V_{j-1}$. Then, the function ψ and the family $\psi_{j,k}(t) = 2^{j/2}\psi(2^j t - k)$, $j, k \in \mathbb{Z}$ are an orthonormal basis of W_j and $L^2(\mathbb{R})$. So that, $L^2(\mathbb{R}) = \oplus_j W_j = V_j \oplus_{k \leq j} W_k$. A signal $f(t) \in L^2(\mathbb{R})$ can be decomposed to [Pou10]

$$f(t) = \sum_{j,k} \langle f(t), \psi_{j,k}(t) \rangle \psi_{j,k}(t)$$

$$= \sum_k \langle f(t), \phi_{j,k}(t) \rangle \phi_{j,k}(t) + \sum_{n,k} \langle f(t), \psi_{n,k}(t) \rangle \psi_{n,k}(t), \tag{10.1}$$

where $n \leq j$. Based on the tight frame property, the expansion coefficients $c(n)$ and $d(n)$ can be obtained by [BGG$^+$98]

$$d(n,k) = \langle f(t), \psi_{n,k}(t) \rangle, \quad \text{and} \quad c(k) = \langle f(t), \phi_{j,k}(t) \rangle. \tag{10.2}$$

Setting these two equations in Eq. 10.1 results in [BGG$^+$98]

$$f(t) = \sum_k c(k)\phi_{j,k}(t) + \sum_{n,k} d(n,k)\psi_{n,k}(t). \tag{10.3}$$

Now computing the sum of the energy of the signal $f(t)$ leads to

$$|f(t)|^2 = |\sum_k c(k)\phi_{j,k}(t) + \sum_{n,k} d(n,k)\psi_{n,k}(t)|^2$$

$$= |\sum_k c(k)\phi_{j,k}(t) \cdot c(k)\phi_{j,k}(t) + \sum_{n,k} d(n,k)\psi_{n,k}(t) \cdot d(n,k)\psi_{n,k}(t)|$$

Based on the orthogonality [Dau90], $\phi_{j,k}(t) \cdot \phi_{j,k}(t) = \delta(k)$ and $\psi_{n,k}(t) \cdot \psi_{n,k}(t) = \delta(k)$. Thus,

$$|f(t)|^2 = |\sum_k c^2(k)\delta(k) + \sum_{n,k} d^2(n,k)\delta(k)|$$

$$= |\sum_k c^2(k) + \sum_{n,k} d^2(n,k)|.$$

Thus, the sum of the energy of the signal $f(t)$ is equal to the sum of its expansion coefficients, iff the basis are orthogonal. $\square$

11 Appendix D

11.1 Equal-length Motif Discovery on Lightning Data Set

The applied Lightning data set in this work comprises three predefined motif types, in which the length of each motif is equal to 318 samples, as illustrated in Fig. 11.1.

In the representation step, the segmented subsequences are decomposed by the ACQTWP transform. Instead of analysing the whole wavelet tree, the bases with the highest amount of information are selected by the early abandoning and BNS algorithms. Table 11.1 represents the results of the two aforementioned algorithms.

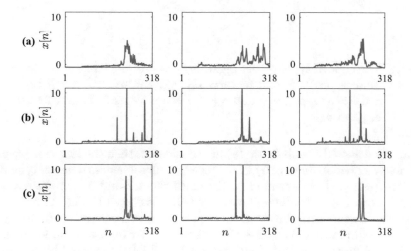

Figure 11.1: (a)-(c) Three types of motifs in the Lightning data set [CKH+16].

According to KITE's structure, six features are extracted from the two selected nodes' coefficients. The performance of these features is analysed by LDA, resulting in the classification error of $0.03 \leq e \leq 0.22$. The entire data cannot be separated linearly since $e \neq 0$. This issue is depicted in Fig. 11.2 (a) and (c).

Table 11.1: Results of the pre-processing step, early abandoning, and Best Nodes Selection algorithms. Pre-processing generates $N = 72$ subsequences with a length of 318 samples. Early abandoning outcomes are scales 1 or 2. The BNS algorithm eventuates the following complex approximation nodes ${}^sQ_{\mathbb{C},1}$, ${}^sQ_{\mathbb{C},2}$, ${}^sQ_{\mathbb{C},5}$ or ${}^sQ_{\mathbb{C},6}$. The selected complex detail nodes are ${}^sR_{\mathbb{C},3}$, ${}^sR_{\mathbb{C},4}$, ${}^sR_{\mathbb{C},7}$, or ${}^sR_{\mathbb{C},8}$.

	Outcomes					
Pre-processing	s_1	s_2	s_3	s_4	...	s_N
Early abandoning	1	2	2	1	...	2
${}^sBQ_{\mathbb{C},i}$	1	2	2	5	...	6
${}^sBR_{\mathbb{C},i}$	4	3	8	4	...	7

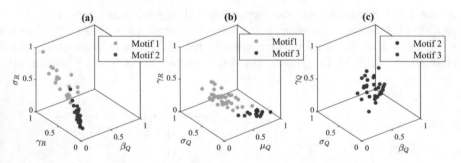

Figure 11.2: Three motif types of the Lightning data set depicted by 3 features.(a) Motif types 1, 2, and features β_Q, γ_R, and σ_R; (b) Motif types 1, 3, and features μ_Q, σ_Q, and γ_R; (c) Motif types 2, 3, and features β_Q, σ_Q, and σ_Q.

If the features depicted in sub-figure (a) are considered, then the two motif types 1 and 2 cannot be divided linearly by LDA since two subsequences of motif type 2 are assigned incorrectly. This is similar to the motif types 2 and 3. On the contrary, motif types 1 and 3 are classified linearly by LDA, as shown in Fig. 11.2 (b).

By measuring the similarity between subsequences' features, motifs are finally detected. The performance of the distance measures under investigation is presented in Table 11.2. Among these tested measures, CD and EDR delivered higher results than other measures.

The LCSS measure achieves the minimum performance due to its sensitivity to noise. Comparing the obtained CR of the tested measures reveals that not all the Lightning data's motifs are detected correctly. The reason is that motif types 2 and 3 are resembling each other, and as illustrated in Fig. 11.2 (a) and (c), some subsequences are assigned to the incorrect motif type. This indicates that the extracted features cannot represent the main characteristics of the data.

The comparison of KITE's performance with state-of-the-art algorithms is provided in Table 11.3.

Table 11.2: Results of equal-length motifs in the Lightning data applying six distance measures.

Distance Measure	Correct motif discovery rate (CR)	Sensitivity (Sn)	Precision (Pr)	F-Measure (F-M)
Euclidean Distance (ED)	0.68	0.67	0.56	0.65
Canberra Distance (CD)	0.77	0.77	0.75	0.76
Dynamic Time Warping (DTW)	0.67	0.67	0.65	0.66
Edit Distance	0.62	0.62	0.60	0.61
Edit Distance on Real Sequence (EDR)	0.71	0.71	0.70	0.71
Longest Common SubSequence (LCSS)	0.52	0.52	0.51	0.51

Table 11.3: Equal-length motif discovery considering different algorithms for the Lightning data set.

Distance Measure	Correct motif discovery rate (CR)	Sensitivity (Sn)	Precision (Pr)	F-Measure (F-M)
KITE	0.77	0.77	0.75	0.76
Brute-Force [PKL+02]	0.68	0.68	0.67	0.67
MOEN Enum. [MuC15]	0.52	0.52	1.00	0.68
MOEN [MuK10]	0.44	0.44	0.43	0.42
Mr.Motif [CaA10]	0.77	0.77	0.65	0.63
SCRIMP++ [ZYZ+18]	0.56	0.56	0.73	0.63
VALMOD [LZP+18]	0.67	0.67	0.61	0.64

The best result is obtained by KITE and Mr.Motif [CaA10] algorithms (77%), and the lowest performance is gained by MOEN [MuK10] (44%). It should be noticed that if motifs of similar types are located in the same sliding window, then the performance of the MOEN algorithm is 100%. The outcomes of the Brute-Force [PKL+02] and VALMOD [LZP+18] methods are similar. The CR achieved by SCRIMP++ [ZYZ+18] is less than 60%. The detected motifs by this method are assigned to the correct type, but its low sensitivity rate (56The MOEN Enum. [MuC15] algorithm could only detect one type of motifs, namely, type 2. Besides this method, other algorithms could determine the three given motif types.

11.2 Variable-length Motif Discovery on Lightning Data Set

In KITE's pre-processing step, the Lightning data set is partitioned into 262 subsequences with lengths vary between 218 and 318 samples. These subsequences are then forwarded to the representation step, which its outcomes are given in Table 11.4.

Table 11.4: Results of the pre-processing step, early abandoning, and Best Nodes Selection algorithms obtained from the Lightning data. The number of $N = 262$ subsequences with lengths between 218 to 318 samples are generated in pre-processing. The outcome of the early abandoning algorithm is the first or second scale. The BNS algorithm detects $^sQ_{\mathbb{C},1}$ or $^sQ_{\mathbb{C},2}$ as the best complex approximation node, and the complex detail node is assigned either by $^sR_{\mathbb{C},3}$ or by $^sR_{\mathbb{C},4}$.

	Outcomes					
Pre-processing	s_1	s_2	s_3	s_4	...	s_N
Early abandoning	1	2	1	1	...	2
$^sBQ_{\mathbb{C},i}$	1	2	2	1	...	1
$^sBR_{\mathbb{C},i}$	4	3	3	3	...	4

The outcome of the early abandoning algorithm is the first or second scale, cf. Table 11.4. The BNS algorithm specifies $^sQ_{\mathbb{C},1}$ or $^sQ_{\mathbb{C},2}$ as the best complex approximation node and $^sR_{\mathbb{C},3}$ or $^sR_{\mathbb{C},4}$ as the best complex detail node.
The effectiveness of the investigated distance measures in the similarity measurement step is provided in Table 11.5.

Table 11.5: Results of variable-length motifs in the Lightning data applying six distance measures.

Distance Measure	Correct motif discovery rate (CR)	Sensitivity (Sn)	Precision (Pr)	F-Measure (F-M)
Euclidean Distance (ED)	0.51	0.51	0.50	0.51
Canberra Distance (CD)	0.73	0.73	0.64	0.68
Dynamic Time Warping (DTW)	0.72	0.72	0.63	0.67
Edit Distance	0.33	0.33	0.45	0.38
Edit Distance on Real Sequence (EDR)	0.73	0.73	0.63	0.68
Longest Common SubSequence (LCSS)	0.26	0.26	0.40	0.32

The maximum CR of 73 % is attained by CD and EDR, and the lowest performance

is achieved by LCSS. DTW provides similar outcomes to CD and EDR.
From all the detected motifs, the first-representative motif pair is illustrated in Fig.
11.3.

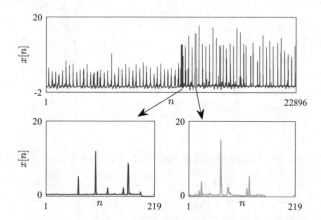

Figure 11.3: Lightning data first-representative motif pair which is detected by
KITE. Both subsequences have equal lengths and they belong to the
motif type 2.

These two subsequences with equal lengths, belonging to motif type 2, are deter-
mined as the first-representative motif from several variable-length detected motifs.
Finally, the results of KITE is compared with MOEN Enum. [MuC15] and VALMOD
[LZP+18] methods in Table 11.6. The lowest performance is achieved by MOEN
Enum. [MuC15] since this method only detects one type of motifs (motif type 1).
KITE and VALMOD [LZP+18] identify all three types of motifs.

Table 11.6: Results of KITE, MOEN Enum. and VALMOD variable-length motifs
for the Lightning data.

Method	Correct motif discovery rate CR	Sensitivity (Sn)	Precision (Pr)	F-Measure (F-M)
KITE	0.73	0.73	0.64	0.68
MOEN Enum. [MuC15]	0.52	0.52	1.00	0.50
VALMOD [LZP+18]	0.64	0.64	0.62	0.63

11.3 KITE's Time Complexity and Distance Measures

KITE's execution time applying CD, DTW, Edit distance, EDR and LCSS for both
synthetic and real-world data is given in the following tables.

Table 11.7: The arithmetic mean value and standard deviation of KITE's execution
time for equal-length motif discovery applying five distance measures.

	Results ($\mu + \sigma$)		
	CD	DTW	Edit D.
Gun Point	1.19+0.01	1.28+0.00	1.80+0.08
50 Words	9.36+0.14	11.88+0.14	31.50+0.56
Lightning	2.86+0.04	4.08+0.13	5.29+0.09
Food	3.50+0.04	3.17+0.02	7.61+0.13
AutoSense-1	96.89+0.16	98.29+0.55	110.41+0.67
AutoSense-2	35.58+0.18	25.67+0.12	35.97+0.38

Table 11.8: The arithmetic mean value and standard deviation of KITE's execution
time for equal-length motif discovery applying five distance measures.

	Results ($\mu + \sigma$)	
	EDR	LCSS
Gun Point	1.23+0.02	2.77+0.02
50 Words	10.81+0.12	68.83+0.60
Lightning	3.01+0.02	9.06+0.04
Food	3.70+0.02	14.02+0.18
AutoSense-1	118.40+3.04	148.99+2.20
AutoSense-2	34.84+0.56	37.85+1.66

Results depicted i tables 11.7, 11.8, 11.9 and 11.10 reveal that the minimum execution time of KITE for all tested data sets is achieved using CD, and the longest
execution time is achieved when LCSS is employed.

The Execution time of KITE for equal-length motif discovery applying CD, DTW,
and EDR is similar. In variable-length motif discovery, the number of subsequences
increases, resulting in higher performance time.

Table 11.9: The arithmetic mean value and standard deviation of KITE's execution time for variable-length motif discovery applying five distance measures.

	Results ($\mu + \sigma$)		
	CD	DTW	Edit D.
Gun Point	7.10+0.03	9.08+0.03	21.62+0.08
50 Words	29.75+0.08	56.08+0.24	234.22+0.07
Lightning	10.84+0.35	14.95+0.07	41.43+0.11
Food	7.95+0.02	10.02+0.01	24.21+0.08
AutoSense-1	395.87+0.11	397.43+2.24	665.66+4.04
AutoSense-2	142.99+1.30	144.12+126	154.31+1.25

Table 11.10: The arithmetic mean value and standard deviation of KITE's execution time for variable-length motif discovery applying five distance
measures.

	Results ($\mu + \sigma$)	
	EDR	LCSS
Gun Point	8.10+0.01	48.76+0.20
50 Words	42.42+0.01	620.42+11.16
Lightning	12.80+0.05	99.24+0.05
Food	9.00+0.08	57.98+0.09
AutoSense-1	392.50+0.44	1574.83+5.90
AutoSense-2	154.21+4.10	200.90+1.35

KITE's execution time for equal- and variable-length motif discovery is benchmarked with state-of-the-art algorithms and demonstrated in Tables 11.11 (Part I)
and 11.12 (Part II).

Table 11.11: The arithmetic mean value and standard deviation of the execution
time of KITE and state-of-the-art algorithms for equal-length motif
discovery - Part I. KITE's results are obtained using ED.

	KITE ($\mu \pm \sigma$)	Mr.Motif ($\mu \pm \sigma$)	MOEN ($\mu \pm \sigma$)
Gun Point	1.24 $\pm$ 0.01	0.85 $\pm$ 0.14	137.98 $\pm$ 0.03
Food	3.81 $\pm$ 0.03	1.43 $\pm$ 0.62	314.92 $\pm$ 0.06
Lightning	3.14 $\pm$ 0.04	1.03 $\pm$ 0.07	359.77 $\pm$ 0.01
50 Words	12.01 $\pm$ 0.46	1.74 $\pm$ 0.24	86.01 $\pm$ 0.01
AutoSense-1	99.73 $\pm$ 4.47	3.36 $\pm$ 0.59	-
AutoSense-2	34.84 $\pm$ 0.56	2.46 $\pm$ 0.54	1096.30 $\pm$ 0.87

Table 11.12: The arithmetic mean value and standard deviation of the execution
time of KITE and state-of-the-art algorithms in equal-length motif
discovery (Part II).

	MOEN Enum. ($\mu \pm \sigma$)	SCRIMP++ ($\mu \pm \sigma$)	VALMOD ($\mu \pm \sigma$)
Gun Point	9.31 $\pm$ 2.04	2.52 $\pm$ 0.08	2.68 $\pm$ 0.11
Food	9.85 $\pm$ 1.86	19.03 $\pm$ 0.09	12.50 $\pm$ 0.15
Lightning	27.80 $\pm$ 0.97	16.13 $\pm$ 0.02	32.91 $\pm$ 0.06
50 Words	1642.60 $\pm$ 1.34	85.71 $\pm$ 1.74	104.45 $\pm$ 0.82
AutoSense-1	-	202.35 $\pm$ 2.29	-
AutoSense-2	40925 $\pm$ 1.24	33553 $\pm$ 1.05	57283.04 $\pm$ 8.83

11.4 KITE's Performance Under Noise

KITE's results obtained from the synthetic data covered with noise (SNR=10dB) and different distance measures for equal- and variable-length motif discovery.

Table 11.13: KITE's equal-length motif discovery applying six distance measures on the Gun data set covered with noise. (EDR: Edit Distance on Real Sequence; lcss:Longest Common SubSequence)

Distance Measure	Correct motif discovery rate (CR)	Sensitivity (Sn)	Precision (Pr)	F-Measure (F-M)
Euclidean Distance	0.78	0.76	0.76	0.76
Canberra Distance	0.78	0.78	0.78	0.78
Dynamic Time Warping	0.87	0.87	0.87	0.87
Edit Distance	0.73	0.72	0.73	0.72
EDRl	0.73	0.73	0.73	0.72
LCSS	0.77	0.77	0.78	0.78

Table 11.14: KITE's equal-length motif discovery applying six distance measures on the 50 words data set covered with noise. (EDR: Edit Distance on Real Sequence; lcss:Longest Common SubSequence)

Distance Measure	Correct motif discovery rate (CR)	Sensitivity (Sn)	Precision (Pr)	F-Measure (F-M)
Euclidean Distance	0.44	0.44	0.43	0.42
Canberra Distance	0.52	0.52	0.50	0.51
Dynamic Time Warping	0.45	0.45	0.44	0.45
Edit Distance	0.43	0.44	0.45	0.44
EDR	0.45	0.45	0.45	0.45
LCSS	0.41	0.41	0.42	0.41

Table 11.15: KITE's equal-length motif discovery applying six distance measures on the Food data set covered with noise. (EDR: Edit Distance on Real Sequence; lcss:Longest Common SubSequence)

Distance Measure	Correct motif discovery rate (CR)	Sensitivity (Sn)	Precision (Pr)	F-Measure (F-M)
Euclidean Distance	0.74	0.74	0.76	0.75
Canberra Distance	0.87	0.87	0.84	0.83
Dynamic Time Warping	0.76	0.76	0.79	0.78
Edit Distance	0.75	0.75	0.73	0.74
EDR	0.89	0.89	0.84	0.86
LCSS	0.72	0.72	0.72	0.71

Table 11.16: KITE's equal-length motif discovery applying six distance measures on the Lightning data set covered with noise.(EDR: Edit Distance on Real Sequence; lcss:Longest Common SubSequence)

Distance Measure	Correct motif discovery rate (CR)	Sensitivity (Sn)	Precision (Pr)	F-Measure (F-M)
Euclidean Distance	0.60	0.61	0.86	0.82
Canberra Distance	0.76	0.76	0.75	0.76
Dynamic Time Warping	0.70	0.70	0.71	0.72
Edit Distance	0.64	0.64	1.00	0.78
EDR	0.76	0.76	0.77	0.76
LCSS	0.68	0.68	0.68	0.68

Table 11.17: KITE's variable-length motif discovery applying six distance measures on the Gun data set covered with noise.(EDR: Edit Distance on Real Sequence; lcss:Longest Common SubSequence)

Distance Measure	Correct motif discovery rate (CR)	Sensitivity (Sn)	Precision (Pr)	F-Measure (F-M)
Euclidean Distance	0.53	0.53	0.54	0.53
Canberra Distance	0.88	0.88	0.85	0.91
Dynamic Time Warping	0.69	0.69	0.69	0.80
Edit Distance	0.55	0.55	0.55	0.70
EDR	0.87	0.87	0.87	0.88
LCSS	0.81	0.81	0.80	0.88

Table 11.18: KITE's variable-length motif discovery applying six distance measures
on the 50 Words data set covered with noise.(EDR: Edit Distance on
Real Sequence; lcss:Longest Common SubSequence)

Distance Measure	Correct motif discovery rate (CR)	Sensitivity (Sn)	Precision (Pr)	F-Measure (F-M)
Euclidean Distance	0.36	0.36	0.35	0.35
Canberra Distance	0.36	0.35	0.35	0.35
Dynamic Time Warping	0.68	0.68	0.65	0.79
Edit Distance	0.33	0.33	0.34	0.35
EDR	0.71	0.71	0.73	0.72
LCSS	0.36	0.36	0.40	0.42

Table 11.19: KITE's variable-length motif discovery applying six distance measures
on the Food data set covered with noise.(EDR: Edit Distance on Real
Sequence; lcss:Longest Common SubSequence)

Distance Measure	Correct motif discovery rate (CR)	Sensitivity (Sn)	Precision (Pr)	F-Measure (F-M)
Euclidean Distance	0.75	0.75	0.75	0.85
Canberra Distance	0.63	0.63	0.64	0.64
Dynamic Time Warping	0.81	0.80	0.81	0.83
Edit Distance	0.51	0.51	0.50	0.51
EDR	0.83	0.87	0.84	0.83
LCSS	0.53	0.53	0.52	0.69

Table 11.20: KITE's variable-length motif discovery applying six distance measures
on the Lightning data set covered with noise.(EDR: Edit Distance on
Real Sequence; lcss:Longest Common SubSequence)

Distance Measure	Correct motif discovery rate (CR)	Sensitivity (Sn)	Precision (Pr)	F-Measure (F-M)
Euclidean Distance	0.45	0.45	0.44	0.44
Canberra Distance	0.62	0.62	0.61	0.74
Dynamic Time Warping	0.55	0.55	0.54	0.55
Edit Distance	0.53	0.53	0.53	0.78
EDR	0.54	0.54	0.55	0.55
LCSS	0.47	0.47	0.46	0.46

11.5 Applications of KITE in Higher Dimension

Besides time series motif discovery in its original form (one-dimensional), it is possible to apply motif discovery on higher-dimensional data. Motif discovery is applicable in image processing applications with various image databases. Image motif discovery aims to detect similar images within an image database without prior information. Such images are called *image motifs*.

Image motifs are detected by KITE applying the same structure. Nevertheless, the pre-processing step (given in Sec. 5.1) is omitted for 2D motif discovery since, unlike methods such as [GSW$^+$14, EPN$^+$15, VDV20], KITE determines two-dimensional image motifs without converting them to one-dimensional signals or segmenting the images. The procedure in the second step is also similar to the one-dimensional data. However, the extension of the ACQTWP transform, namely, the 2D-ACQTWP transform, must be employed, explained in the next section. Finally, employing the feature extraction and similarity measurement steps result in detecting image motifs.

After introducing the 2D-ACQTWP transform, KITE's performance on a synthetic image data set is examined. It should be noted that the work of the author published in [ToL17a, ToL18] is integrated literally into this section.

11.5.1 Two Dimensional Analytic Complex Quad Tree Wavelet Packet

The first scale of the 2D-ACQTWP is similar to the 2D-discrete wavelet transform (2D-DWT) [BGG$^+$98]. In the first scale, 2D-ACQTWP decomposes an image into four sub-bands. This is the same for each wavelet packets A and B (2D-WPT A, 2D-WPT B) in the first scale. The structure of two scales decomposition of the 2D-WPT A is depicted in Fig. 11.4(b), where both low and high-pass filtered sub-bands decomposed further.

The sub-band LL_1 in the first scale is obtained by low-pass filtering the input image along the first dimension (row) and then again low-pass filtering it along the second dimension (column) [1]. Low-pass filtering the image along the first dimension and then high-pass filtering it along the second dimension in the first scale result in LH_1. Similarly, the HL_1 and HH_1 are labelled, and index 1 determines the decomposed scale. The same procedure is performed on each sub-band in order to obtain the second scale coefficients.

Thus, each sub-band of the 2D-ACQTWP transform is a product of scaling and wavelet functions along an image's rows and columns.

The definition of these functions is based on the definition of one-dimensional scaling and wavelet functions of the ACQTWP transform. For simplicity, the wavelet

---

[1] An image X is given by a matrix of size $n \times m$, where $n, m \in \mathbb{N}$ are the number of rows and columns.

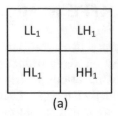

(a)

LL₁LL₂	LL₁LH₂	LH₁LL₂	LH₁LH₂
LL₁HL₂	LL₁LH₂	LH₁HL₂	LH₁HH₂
HL₁LL₂	HL₁LH₂	HH₁LL₂	HH₁LH₂
HL₁HL₂	HL₁HH₂	HH₁HL₂	HH₁HH₂

(b)

Figure 11.4: Structure of two scales decomposition of the 2D-WPT A: (a) the first scale decomposed an image into four sub-bands, (b) the second scale decomposition results in 16 sub-bands, each sub-band obtains by first filtering the image along rows (first dimension) and then along columns (second dimension) [ToL18].

and scaling functions of WPT A defined in Def. 5.5 are denoted by

$$\psi_{A,e}(t) = \psi_{A,2J+1}(t), \qquad \psi_{A,o}(t) = \psi_{A,2J+3}(t);$$
$$\phi_{A,e}(t) = \phi_{A,2J}(t), \qquad \phi_{A,o}(t) = \phi_{A,2J+2}(t).$$

The indices e and o denote the even and odd sub-sampled functions. The functions of WPT B are represented in the same manner. By applying this new notation, the wavelet and scaling functions of the 2D-ACQTWP are defined as:

Definition 11.1 (2D-ACQTWP scaling and wavelet functions). Twelve wavelets and four scaling functions are employed in 2D-WPT A. The wavelet functions are given by [ToL18]

$$\psi_{A,1}(x, y) = \phi_{A,e}(x)\psi_{A,e}(y), \qquad \psi_{A,4}(x, y) = \phi_{A,e}(x)\psi_{A,o}(y),$$

$$\psi_{A,2}(x, y) = \psi_{A,e}(x)\phi_{A,e}(y), \qquad \psi_{A,5}(x, y) = \psi_{A,e}(x)\phi_{A,o}(y),$$

$$\psi_{A,3}(x, y) = \psi_{A,e}(x)\psi_{A,e}(y), \qquad \psi_{A,6}(x, y) = \psi_{A,e}(x)\psi_{A,o}(y),$$

where the 2D-wavelet $\psi(x, y) = \psi(x)\psi(y)$ represents the row-column implementation of the wavelet transformation. The rest of the wavelet functions are obtained in the similar procedure. The four scaling functions are defined as [ToL18]

$$\phi_{A,1}(x,y) = \phi_{A,e}(x)\phi_{A,e}(y), \qquad \phi_{A,2}(x,y) = \phi_{A,e}(x)\phi_{A,o}(y),$$

$$\phi_{A,3}(x,y) = \phi_{A,o}(x)\phi_{A,e}(y), \qquad \phi_{A,4}(x,y) = \phi_{A,o}(x)\phi_{A,o}(y).$$

The wavelet and scaling functions of the 2D-WPT B are given accordingly [ToL18].

The outcomes of the 2D-ACQTWP transform are complex coefficients, obtained only when the coefficients of both wavelet packet trees are considered. The wavelet and scaling coefficients of the 2D-ACQTWP for the 2D-WPT A are given by

Definition 11.2 (2D-ACQTWP's coefficients of 2D-WPT A). Coefficients of the 2D-ACQTWP for the 2D-WPT A are denoted by ${}^sC[x,y] = \{\ {}^{s+1}C_{2J}[x,y],\ {}^{s+1}C_{2J+1}[x,y], ...,\ {}^{s+1}C_{2J+15}[x,y]\}$. Parameter $s \in \mathbb{N}$ is number of scales and $J = 8j$ is the index of the nodes, whereby for the first scale $s = 1$, $j = 0$ and for scales greater than one, $s > 1$ $0 \le j < 4^s$. These coefficients are obtained as follows [ToL18]:

$$^{s+1}C_{2J}[x,y] = \ ^sC_j[x,y] * \ ^sf_1[2x] \ ^sf_1[2y],$$

$$^{s+1}C_{2J+1}[x,y] = \ ^sC_j[x,y] * \ ^sf_1[2x+1] \ ^sf_1[2y],$$

$$\vdots$$

$$^{s+1}C_{2J+5}[x,y] = \ ^sC_j[x,y] * \ ^sf_1[2x] \ ^sf_2[2y],$$

$$^{s+1}C_{2J+6}[x,y] = \ ^sC_j[x,y] * \ ^sf_1[2x] \ ^sf_2[2y+1],$$

$$\vdots$$

$$^{s+1}C_{2J+14}[x,y] = \ ^sC_j[x,y] * \ ^sf_2[2x] \ ^sf_2[2y],$$

$$^{s+1}C_{2J+15}[x,y] = \ ^sC_j[x,y] * \ ^sf_2[2x+1] \ ^sf_2[2y+1].$$

The filters f_1 and f_2 are obtained by:

$$(f_1,f_2) = \begin{cases} (g_a, h_a) & j = 0 \text{ (first branch)}, \\ (h_b, g_b), & j = (2^s \cdot (s-1)) - 1 \text{ (last branch)}, \\ \text{Swap the filters in each} & \\ \text{branchwith respect to} & \text{rest.} \\ \text{the previous branch.} & \end{cases}$$

Filters ${}^sh_{a,b}$ and ${}^sg_{a,b}$ are high and low-pass filters in wavelet packets A and B.

Definition 11.3 (2D-ACQTWP's coefficients of 2D-WPT B). The wavelet coefficients for the 2D-WPT B are computed similarly and denoted by

$$^sD[x,y] = \{\ {}^{s+1}D_{2J}[x,y],\ {}^{s+1}D_{2J+1}[x,y], ...,\ {}^{s+1}D_{2J+15}[x,y]\}.$$

Definition 11.4 (2D-ACQTWP complex coefficients). Complex coefficients of the 2D-ACQTWP transform are denoted by ${}^s\mathbb{C}$, and obtained by: ${}^s\mathbb{C}[x,y] =$

${}^sC[x, y] + i\ {}^sD[x, y]$, where $s \in \mathbb{N}$ is the number of scales, $i^2 = -1$, and ${}^sC[x, y]$ and ${}^sD[x, y]$ are coefficients of 2D-WPT A and B, respectively.

Similar to the ACQTWP transform, the coefficients of the 2D-ACQTWP transform as well can be classified into approximation and detail coefficients.

Definition 11.5 (Approximation and detail coefficients of 2D-ACQTWP). The coefficients of 2D-ACQTWP for the wavelet packet A in scale s are divided into approximation (scaling) ${}^sQ_A[n]$ and detail (wavelet) ${}^sR_A[n]$ coefficients by

$$
\begin{aligned}
{}^sQ_{A,l}[n] &= \{{}^sC_{2J}[n],\ {}^sC_{2J+1}[n],\ {}^sC_{2J+2}[n],\ {}^sC_{2J+3}[n]\}, \\
{}^sR_{A,k}[n] &= \{{}^sC_{2J+4}[n],\ {}^sC_{2J+5}[n],\ {}^sC_{2J+6}[n], ...,\ {}^sC_{2J+15}[n]\},
\end{aligned}
\tag{11.1}
$$

where in the first scale $s = 1$, $l = 1$ and $k = \{2, 3, 4\}$. In scales greater than one ($s > 1$), $l = \{1, 2, 3, 4\}$ and $k = j - 4$ for $5 \le j < 4^s$. For the 2D-WPT B the approximation and detail coefficients are obtained similarly, and denoted by ${}^sQ_{B,l}[n]$ and ${}^sR_{B,k}[n]$.

Thus, for $s = 1$, the approximation coefficients of the 2D-WPT A are given by ${}^1Q_{A,1}[n] = \{{}^sC_{2J}[n],\ {}^sC_{2J+1}[n],\ {}^sC_{2J+2}[n],\ {}^sC_{2J+3}[n]\}$ and the detail coefficients are obtained by

$$
{}^1R_{A,2}[n] = \{{}^sC_{2J+4}[n],\ {}^sC_{2J+5}[n],\ {}^sC_{2J+6}[n],\ {}^sC_{2J+7}[n]\},
$$

$$
{}^1R_{A,3}[n] = \{{}^sC_{2J+8}[n],\ {}^sC_{2J+9}[n],\ {}^sC_{2J+10}[n],\ {}^sC_{2J+11}[n]\},
$$

$$
{}^1R_{A,4}[n] = \{{}^sC_{2J+12}[n],\ {}^sC_{2J+13}[n],\ {}^sC_{2J+14}[n],\ {}^sC_{2J+15}[n]\}.
$$

The complex approximation and detail coefficients considering both wavelet packet trees are denoted by ${}^sQ_{C,l}$ and ${}^sR_{C,k}$, respectively.

The 2D-ACQTWP transform has the same properties of the one-dimensional ACQTWP transform.

11.5.1.1 Motif Discovery on Image Data Set

In order to analyse the performance of KITE, a synthetic image data set is provided, which comprises images from diverse applications like hand gestures, leaf identification [SMRA14], and text and object recognition, as represented in Fig. 11.5 [ToL18].

This data set consists of 2202 images of various sizes, of which 400 images are labelled as motif images. First, all of the images are converted to grey-scale images since this application focuses on analysing KITE's performance when the colour information is not available [ToL18]. Next, these images are forwarded to the representation step and investigated by the 2D-ACQTWP transform. The decomposition stops at the third scale, and the best scale is given by $(s_b = \{2, 3\})$. The BNS algorithm detects the two best complex nodes (one approximation and one detail node) for feature extraction.

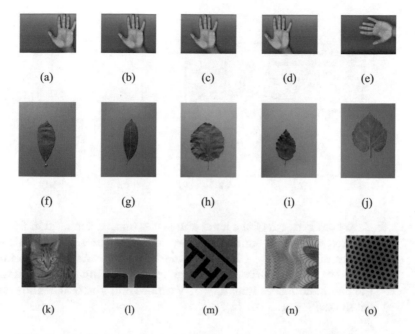

Figure 11.5: A synthetic data set of different images with various sizes gathered
from various applications such as hand gesture, leaf, and object recognition.

Like the one-dimensional motif discovery, the selected-nodes' coefficients must be
normalised to provide an ability to compare images of different sizes. The normalisation is performed by the normalised histogram obtained from the absolute value
of the complex coefficients of the selected-nodes.

In addition to the first four statistical moments from the normalised histograms, the
energy of the wavelet coefficients are also extracted as features. As already stated,
the shape characteristics of the normalised histograms are determined by the first
four moments. Due to shift-invariant property of the 2D-ACQTWP transform,
the coefficients of a signal and its shifted versions are identical so, the amount of
their energy is equal [ToL18]. Additionally, according to the orthonormality of
the scaling and wavelet functions of 2D-ACQTWP and Parseval's theorem 3, the
enrgy of the signal is preserved [ToL18]. Thus, energy of the wavelet coefficients is
included as features.

The efficiency of the extracted features are measured by LDA algorithm [Fis36,
Alp10], resulting in the classification error of $0 \leq c \leq 0.01$. Figure 11.6 illustrates
the performance of these features for two groups of images [ToL18].

Both of these motif groups can be separated linearly. The distance between features of the same motif group (represented on the projection line) is minimised.
Moreover, the features of the hand image and its shifted versions are all mapped
to the same point. The reason for that is the 2D-ACQTWP transform, as well as

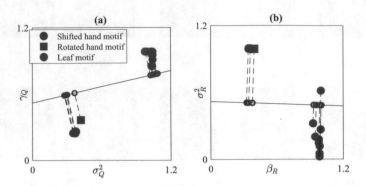

Figure 11.6: Extracted features of the hand and leaves images depicted in Fig. 11.5. The red circle and square markers represent the shifted and rotated images of the hand. Blue circle markers demonstrate leaves images. (a) Skewness and variance of the complex approximation coefficients of the hand and leaves images. (b) Variance and kurtosis of the complex detail coefficients [ToL18].

the normalised histograms, are shift invariant [ToL18]. Finally, image motifs are detected after quantifying their similarity. The results of this step, applying six distance measures, are given in Table 11.21.

Table 11.21: KITE 2D variable size motif discovery applying six distance measures.

Distance Measure	Correct motif discovery rate (CR)	Sensitivity (Sn)	Precision (Pr)	F-Measure (F-M)
Euclidean Distance	0.81	0.81	0.82	0.82
Canberra Distance	0.93	0.93	0.91	0.92
Dynamic Time Warping	0.81	0.81	0.79	0.80
Edit Distance	0.62	0.62	0.60	0.61
EDR	0.62	0.62	0.61	0.63
LCSS	0.34	0.34	0.34	0.34

CD obtains the highest outcomes, and LCSS provides the lowest results due to its noise sensitivity. From 400 image motifs, the CD detects 370 image motifs. The ED and DTW obtain the next best results. Each of these measures identifies, respectively, 327 and 322 image motifs. The Edit distance and EDR distinguish 244 and 154 motifs. Finally, the number of 139 image motifs are determined by the LCSS measure.

To analyse KITE's robustness against noise, the test cases are covered with the Gaussian and Salt & Pepper noise [Bon09]. The added Gaussian and Salt & Pepper noises are, SNR=20dB and SNR=13dB, respectively. KITE's performance under the influence of noise is provided in Tables 11.22 - 11.23. Due to its noise sensitivity, LCSS obtains a CR under 50 %, and CD outperforms the other measures. The rest of the distance measures perform alike. The CR obtained from the noisy test data is lower than the data without noise. Besides noise, another occurring distortion is image blurring [BuB09]. KITE's outcomes on blurred images are given in Table 11.24. The highest CR is obtained by the DTW measure (CR=0.81) and the lowest rate by LCSS (CR=0.33). The ED obtains higher results than CD since CD is very sensitive to the values close to zero [ToL18].

The performance of KITE in 2D-variable size motif discovery cannot be benchmarked with other state-of-the-art algorithms [ToL18] since, on the contrary to other algorithms [CFC+12, GSW+14, EPN+15], KITE does not convert images to one-dimensional data.

Table 11.22: KITE's results on images overlaid with Gaussian noise.

Distance Measure	Correct motif discovery rate (CR)	Sensitivity (Sn)	Precision (Pr)	F-Measure (F-M)
Euclidean Distance	0.78	0.78	0.77	0.77
Canberra Distance	0.83	0.83	0.82	0.83
Dynamic Time Warping	0.78	0.78	0.77	0.77
Edit distance	0.59	0.59	0.58	0.59
EDR	0.60	0.59	0.59	0.60
LCSS	0.33	0.33	0.32	0.33

Table 11.23: KITE's results on images overlaid with Salt & Pepper noise.

Distance Measure	Correct motif discovery rate (CR)	Sensitivity (Sn)	Precision (Pr)	F-Measure (F-M)
Euclidean Distance	0.73	0.73	0.72	0.73
Canberra Distance	0.83	0.83	0.82	0.82
Dynamic Time Warping	0.75	0.75	0.74	0.74
Edit distance	0.56	0.56	0.56	0.55
EDR	0.53	0.53	0.52	0.52
LCSS	0.33	0.33	0.35	0.34

Table 11.24: Results of KITE and six distance measures for blurred images.

Distance Measure	Correct motif discovery rate (CR)	Sensitivity (Sn)	Precision (Pr)	F-Measure (F-M)
Euclidean Distance	0.80	0.80	0.80	0.79
Canberra Distance	0.78	0.79	0.77	0.78
Dynamic Time Warping	0.81	0.81	0.79	0.80
Edit distance	0.52	0.52	0.51	0.52
EDR	0.53	0.53	0.53	0.53
LCSS	0.33	0.33	0.32	0.33

Bibliography

[AAJ+19]   ALI, M. ; ALQAHTANI, A. ; JONES, M. W. ; XIE, X.: Clustering and
            Classification for Time Series Data in Visual Analytics: A Survey.
            In: *IEEE Access* 7 (2019), P. 181314–181338

[AbS04]   ABDELNOUR, F. A. ; SELESNICK, I.:   Symmetric nearly shift-
            invariant tight frame wavelets. In: *IEEE Transactions on signal
            processing* 53 (2004), Nr. 1, P. 231–239. – IEEE

[Ada10]   ADAM, I.: *Complex Wavelet Transform: application to denoising.*
            Editura Politehnică, 2010. – PhD Thesis, Politehnica University of
            Timisoara and Université de Rennes

[AhR12]   AHMED, N. ; RAO, K. R.: *Orthogonal transforms for digital signal
            processing.* 1st ed. Springer Science & Business Media, 2012. – ISBN
            978–3–642–45452–3

[AlA20]   AL AGHBARI, Z. ; AL-HAMADI, A.: Finding K Most Significant
            Motifs in Big Time Series Data. In: *Procedia Computer Science* 170
            (2020), P. 595–601

[AlD94]   ALUR, R. ; DILL, D. L.: A theory of timed automata. In: *Theoretical
            computer science* 126 (1994), Nr. 2, P. 183–235. – Elsevier

[Alp10]   ALPAYDIN, E.: *Introduction to machine learning.* 2nd. ed. Cam-
            bridge : The MIT Press, 2010. – ISBN 9780262012430

[AnA16]   ANTONINO, V. O. ; ARAUJO, A. F. R.: Local adaptive receptive field
            dimension selective self-organizing map for multi-view clustering. In:
            International Joint Conference on Neural Networks, 2016, P. 698–
            705. – IEEE

[Anv16]   ANH, D. T. ; VAN NHAT, N.: An efficient implementation of EMD
            algorithm for motif discovery in time series data. In: *International
            Journal of Data Mining, Modelling and Management* 8 (2016), Nr.
            2, P. 180. – Inderscience Publishers

[ASC19]   ATTALLAH, B. ; SERIR, A. ; CHAHIR, Y.: Feature extraction in
            palmprint recognition using spiral of moment skewness and kurtosis
            algorithm. In: *Pattern Analysis and Applications* 22 (2019), Nr. 3,
            P. 1197–1205. – Springer

© The Editor(s) (if applicable) and The Author(s), under exclusive license to
Springer-Verlag GmbH, DE, part of Springer Nature 2022
S. Deppe, *Discovery of Ill-Known Motifs in Time Series Data*, Technologien
für die intelligente Automation 15, https://doi.org/10.1007/978-3-662-64215-3

[ASS+15]   AGARWAL, P. ; SHROFF, G. ; SAIKIA, S. ; KHAN, Z.: Efficiently dis-
             covering frequent motifs in large-scale sensor data. In: *Proceedings
             of the Second IKDD Conference on Data Sciences*, ACM, 2015, P.
             98–103

[Aus93]   AUSCHER, P.: Compactly supported wavelets and boundary con-
             ditions. In: *Journal of Functional Analysis* 111 (1993), Nr. 1, P.
             29–43. – Elsevier

[BaM88]   BALANDA, K. P. ; MACGILLIVRAY, H. L.: Kurtosis: a critical
             review. In: *The American Statistician* 42 (1988), Nr. 2, P. 111–119.
             – Taylor & Francis Group

[BaP66]   BAUM, L. E. ; PETRIE, T.: Statistical inference for probabilistic
             functions of finite state markov chains, Institute of Mathematical
             Statistics, 1966, P. 1554–1563

[BaR16]   BAYDOGAN, M. ; RUNGER, G.: Time series representation and sim-
             ilarity based on local autopatterns. In: *Data Mining and Knowledge
             Discovery* 30 (2016), Nr. 2, P. 476–509. – Springer

[BaS08]   BAYRAM, I. ; SELESNICK, I. W.:   On the Dual-Tree Complex
             Wavelet Packet and M -Band Transforms. In: *IEEE Transactions
             on Signal Processing* 56 (2008), Nr. 6, P. 2298–2310. – IEEE

[BaS10]   BAR, B. ; SAPIRO, G.: Hierarchical dictionary learning for invari-
             ant classification. In: *IEEE International Conference on Acoustics,
             Speech and Signal Processing*, IEEE, 2010, P. 3578–3581

[BBB+09]   BAILEY, T. L. ; BODEN, M. ; BUSKE, F. A. ; FRITH, M. ; GRANT,
             C. E. ; CLEMENTI, L. ; REN, J. ; LI, W. W. ; NOBLE, W. S.:
             MEME Suite: tools for motif discovery and searching. In: *Nucleic
             Acids Research* 37 (2009), Nr. suppl-2, P. W202–W208

[BDD+19]   BATOR, M. ; DICKS, A. ; DEPPE, S. ; LOHWEG, V.: Anomaly De-
             tection with Root Cause Analysis for Bottling Process. In: *24nd In-
             ternational Conference on Emerging Technologies and Factory Au-
             tomation (ETFA)*, IEEE, 2019, P. 1619–1622

[BDH+12]   BAGNALL, A. ; DAVIS, l. ; HILLS, J. ; LINES, J.: Transformation
             Based Ensembles for Time Series Classification. In: *Proceedings of
             the 12th SIAM International Conference on Data Mining*, 2012, P.
             307–319. – SIAM

[BDM+12]   BATOR, M. ; DICKS, A. ; MÖNKS, U. ; LOHWEG, V.:   Feature
             extraction and reduction applied to sensorless drive diagnosis. In:
             22. Workshop Computational Intelligence Bd. 45. Karlsruhe : KIT

Scientific Publishing, 2012 (Schriftenreihe des Instituts für Angewandte Informatik - Automatisierungstechnik am Karlsruher Institut für Technologie), P. 163–178

[BeC94]   BERNDT, D. J. ; CLIFFORD, J.: Using dynamic time warping to find patterns in time series. In: *Knowledge Discovery and Data Mining workshop*, AAAI Press, 1994 (10 No. 16), P. 359–370

[Bed62]   BEDROSIAN, E.: A product theorem for Hilbert transforms. (1962)

[BeK18]   BELOV, A. A. ; KROPOTOV, Y. A.: Automated Collection and Processing Data in the System of Monitoring Environmental Pollution by Industrial Enterprises. In: *International Multi-Conference on Industrial Engineering and Modern Technologies*, IEEE, 2018, P. 1–5

[Ber09]   BERGER, M.: *Geometry I.* 3rd ed. Springer Science & Business Media, 2009. – ISBN 978–3–540–11658–5

[BeR14]   BETTAIAH, V. ; RANGANATH, H. S.: An Analysis of Time Series Representation Methods: Data Mining Applications Perspective. In: *Proceedings of the 2014 ACM Southeast Regional Conference*, ACM, 2014 (ACM SE '14), P. 16:1–16:6

[Bev12]   BERLIN, E. ; VAN LAERHOVEN, K.: Detecting leisure activities with dense motif discovery. In: *Proceedings of the ACM Conference on Ubiquitous Computing*, ACM, 2012 (UbiComp '12), P. 250–259

[BGG+98]   BURRUS, C. S. ; GOPINATH, R. A. ; GUO, H. ; ODEGARD, J. E. ; SELESNICK, I. W.: *Introduction to wavelets and wavelet transforms: a primer.* Bd. 23. Prentice Hall Upper Saddle River, 1998. – ISBN 978–0134896007

[BHW+04]   BALASUBRAMANIYAN, R. ; HÜLLERMEIER, E. ; WESKAMP, N. ; KÄMPER, J.: Clustering of gene expression data using a local shape-based similarity measure. In: *Bioinformatics* 21 (2004), Nr. 7, P. 1069–1077. – Oxford University Press

[Bis07]   BISHOP, C. M.: *Pattern Recognition and Machine Learning (Information Science and Statistics).* 1st ed. 2006. Corr. 2nd printing. Springer-Verlag, New York, Inc., Secaucus, NJ, 2007. – ISBN 0387310738

[BJR+15]   BOX, G. E. P. ; JENKINS, G. M. ; REINSEL, G. C. ; LJUNG, G. M.: *Time series analysis: forecasting and control.* 5th ed. John Wiley & Sons, 2015. – ISBN 978–1–118–67502–1

[BKT+14]   BATISTA, G. E. A. P. A. ; KEOGH, E. ; TATAW, O. M. ; SOUZA,
            V. M. A.: CID: an efficient complexity-invariant distance for time
            series. In: *Data Mining and Knowledge Discovery* 28 (2014), Nr. 3,
            P. 634–669. – Springer

[BLB+17]   BAGNALL, A. ; LINES, J. ; BOSTROM, A. ; LARGE, J. ; KEOGH, E.:
            The great time series classification bake off: a review and experi-
            mental evaluation of recent algorithmic advances. In: *Data Mining
            and Knowledge Discovery* 31 (2017), Nr. 3, P. 606–660. – Springer

[Blu09]   BLUMAN, A. G.:   *Elementary statistics: A step by step ap-
            proach.* 10th ed. McGraw-Hill Higher Education, 2009. – ISBN
            9781259755330

[Bon09]   BONCELET, C.: Image noise models. In: *The Essential Guide to
            Image Processing.* Elsevier, 2009, P. 143–167

[BPC+18]   BENTO, P. M. R. ; POMBO, J. ; CALADO, M. R. A. ; MARIANO,
            S.J. P. S.: A bat optimized neural network and wavelet transform
            approach for short-term price forecasting. In: *Applied energy* 210
            (2018), P. 88–97. – Elsevier

[BrB74]   BRIGHAM, E. O. ; BRIGHAM, E. O.: *The fast Fourier transform.*
            Bd. 7. Prentice-Hall Englewood Cliffs, NJ, 1974. – ISBN 0–13–
            307505–2

[BrB18]   BRITO, R. C. ; BASSANI, H.F.: Self-Organizing Maps with Variable
            Input Length for Motif Discovery and Word Segmentation. In: *In-
            ternational Joint Conference on Neural Networks*, 2018, P. 1–8. –
            IEEE

[BRL04]   BURKHARDT, H. ; REISERT, M. ; LI, H.:   Invariants for Dis-
            crete Structures–An Extension of Haar Integrals over Transforma-
            tion Groups to Dirac Delta Functions. In: *Joint Pattern Recognition
            Symposium*, Springer, 2004, P. 137–144

[BrM80]   BRACCINI, C. ; MARINO, g.:   Fast geometrical manipulations of
            digital images. In: *Computer Graphics and Image Processing* 13
            (1980), Nr. 2, P. 127–141

[BSM+01]   BRONSTEIN, I. N. ; SEMENDJAJEW, K. A. ; MUSIOL, G. ; MÜHLIG,
            H. ; AUFLAGE 5. überarbeitete und e. (Hrsg.): *Taschenbuch der
            Mathematik.* Frankfurt am Main : Verlag Harri Deutsch, 2001. –
            ISBN 3–8171–2015–X

[BuB09]   BURGER, W. ; BURGE, M. J.: *Principles of digital image processing.*
            Springer-Verlag London, 2009. – ISBN 978–1–84800–190–9

[BuK15]   BUTLER, M. ; KAZAKOV, D.: Sax discretization does not guarantee
            equiprobable symbols. In: *IEEE Transactions on Knowledge and
            Data Engineering* 27 (2015), Nr. 4, P. 1162–1166. – IEEE

[BuL17]   BUDUMA, N. ; LOCASCIO, N.: *Fundamentals of deep learning:
            Designing next-generation machine intelligence algorithms.* 1st ed.
            "O'Reilly Media, Inc.", 2017. – ISBN 978–1491925614

[BuM80]   BURKHARDT, H. ; MÜLLER, X.: On invariant sets of a certain class
            of fast translation-invariant transforms. In: *IEEE Transactions on
            Acoustics, Speech and Signal Processing* 28 (1980), Nr. 5, P. 517–
            523. – IEEE

[BuS95]   BURKHARDT, H. ; SCHULZ-MIRBACH, H.: A contribution to non-
            linear system theory. In: *Proc. of the IEEE Workshop on Nonlinear
            Signal and Image Processing* Bd. 2, IEEE, 1995, P. 823–826

[BuT01]   BUHLER, J. ; TOMPA, M.: Finding motifs using random projections.
            In: *Proceedings of the Fifth Annual International Conference on
            Computational Biology*, ACM, 2001 (RECOMB '01), P. 69–76

[BWP16]   BALASUBRAMANIAN, A. ; WANG, J. ; PRABHAKARAN, B.: Discov-
            ering multidimensional motifs in physiological signals for personal-
            ized healthcare. In: *Journal of Selected Topics in Signal Processing*
            (2016), Nr. 99, P. 1. – IEEE

[CaA93]   CAGLAR, H. ; AKANSU, A. N.: A generalised parametric PR-QMF
            design technique based on Bernstein polynomial approximation. In:
            IEEE Transactions on Signal Processing 41 (1993), Nr. 7, P. 2314–
            2321. – IEEE

[CaA10]   CASTRO, N. ; AZEVEDO, P. J.: Multiresolution motif discovery in
            time series. In: *Proceedings of the International Conference on Data
            Mining*, SIAM, 2010, P. 665–676

[CaL17]   CALUDE, C. S. ; LONGO, G.: The Deluge of Spurious Correlations
            in Big Data. In: *Foundations of science* 22 (2017), Nr. 3, P. 595–612

[CBA16]   COSENTINO, R. ; BALESTRIERO, R. ; AAZHANG, B.: Best ba-
            sis selection using sparsity driven multi-family wavelet transform.
            In: *IEEE Global Conference on Signal and Information Processing*,
            IEEE, 2016, P. 252–256

[CCB04]   CHAN, W. L. ; CHOI, H. ; BARANIUK, R.: Quaternion wavelets
            for image analysis and processing. In: *International Conference on
            Image Processing* Bd. 5, IEEE, 2004, P. 3057–3060

[CFC+12]   CHI, L. ; FENG, Y. ; CHI, H. ; HUANG, Y.: Face image recogni-
            tion based on time series motif discovery. In: *IEEE International
            Conference on Granular Computing*, IEEE, 2012, P. 72–77

[Che05]   CHEN, L.: *Similarity search over time series and trajectory data*, University of Waterloo, Diss., 2005

[Che09]   CHEN, W.-K.: *Passive, active, and digital filters.* 3th ed. Crc Press, 2009. – ISBN 1420058851. – Series: The Circuits and Filters Handbook

[Ciz70]   CIZEK, V.: Discrete Hilbert transform. In: *IEEE Transactions on Audio and Electroacoustics* 18 (1970), Nr. 4, P. 340–343. – IEEE

[CKH+16]   CHEN, Y. ; KEOGH, E. ; HU, B. ; BEGUM, N. ; BAGNALL, A. ; MUEEN, A. ; BATISTA, G.: *The UCR Time Series Classification Archive.* 2016. – www.cs.ucr.edu/~eamonn/time_series_data/; Last access: 19.01.2020.

[CKL03]   CHIU, B. ; KEOGH, E. ; LONARDI, S.: Probabilistic discovery of time series motifs. In: *Proceedings of the Ninth ACM SIGKDD International Conference on Knowledge Discovery and Data Mining,* ACM, 2003 (KDD '03), P. 493–498

[CLS+10]   CALONDER, M. ; LEPETIT, V. ; STRECHA, C. ; FUA, P.: Brief: Binary robust independent elementary features. In: *European conference on computer vision,* Springer, 2010, P. 778–792

[CLZ17]   CHAI, P. ; LUO, X. ; ZHANG, Z.: Image Fusion Using Quaternion Wavelet Transform and Multiple Features. In: *IEEE Access* 5 (2017), P. 6724–6734. – IEEE

[CNO+07]   CHEN, Y. ; NASCIMENTO, M. A. ; OOI, B. C. ; TUNG, A.: SpADe: On shape-based pattern detection in streaming time series. In: *IEEE 23rd International Conference on Data Engineering,* IEEE, 2007, P. 786–795

[COO05]   CHEN, L. ; ÖZSU, M. T. ; ORIA, V.: Robust and Fast Similarity Search for Moving Object Trajectories. In: *Proceedings of the 2005 ACM SIGMOD International Conference on Management of Data,* Association for Computing Machinery, 2005 (SIGMOD '05), P. 491–502

[CoR95]   COHEN, A. ; RYAN, R. D.: *Wavelets and multiscale signal processing.* Springer, 1995. – ISBN 978–0–412–57590–7

[CoV95]   CORTES, C. ; VAPNIK, V.: Support-vector networks. In: *Machine learning* 20 (1995), Nr. 3, P. 273–297. – Springer

[Cra10]   CRAW, S.: *Manhattan Distance.* Springer US, 2010. – 639–639 P. – ISBN 978–0–387–30164–8

[CRF+19]   CARRERA, D ; ROSSI, B ; FRAGNETO, P ; BORACCHI, G: Online anomaly detection for long-term ecg monitoring using wearable devices. In: *Pattern Recognition* 88 (2019), P. 482–492. – Elsevier

[CST+00]   CRISTIANINI, N. ; SHAWE-TAYLOR, J. u. a.:   *An introduction to support vector machines and other kernel-based learning methods.* Cambridge university press, 2000. – ISBN 9780521780193

[CTC+19]   CANIZO, M. ; TRIGUERO, I. ; CONDE, A. ; ONIEVA, E.: Multi-head CNN–RNN for multi-time series anomaly detection: An industrial case study. In: *Neurocomputing* 363 (2019), P. 246–260

[DaD07]   DAS, M. K. ; DAI, H.-K.: A survey of DNA motif finding algorithms. In: *BMC bioinformatics* 8 (2007), Nr. 7, P. 21

[Dau90]   DAUBECHIES, I.:   The wavelet transform, time-frequency localization and signal analysis. In: *IEEE Transactions on Information Theory* 36 (1990), Nr. 5, P. 961–1005. – IEEE

[DeC97]   DECARLO, L. T.: On the meaning and use of kurtosis. In: *Psychological methods* 2 (1997), Nr. 3, P. 292. – American Psychological Association

[DeD09]   DEZA, M. M. ; DEZA, E.:   *Encyclopedia of distances.*   Springer, Berlin, Heidelberg, 2009. – ISBN 978–3–642–00233–5

[DLC+19]   DU, J. ; LIU, Q. ; CHEN, K. ; WANG, J.: Forecasting stock prices in two ways based on LSTM neural network. In: *IEEE 3rd Information Technology, Networking, Electronic and Automation Control Conference*, IEEE, 2019, P. 1083–1086

[DLW+15]   DICKS, A. ; LOHWEG, V. ; WITTKE, H. ; LINKE, S.: Structural health monitoring of plastic components with piezoelectric sensors. In: *20th Conference on Emerging Technologies and Factory Automation (ETFA)*, IEEE, 2015, P. 1–4

[DSP+17]   DAU, H. A. ; SILVA, D. F. ; PETITJEAN, F. ; FORESTIER, G. ; BAGNALL, A. ; KEOGH, E.: Judicious setting of Dynamic Time Warping's window width allows more accurate classification of time series. In: *IEEE International Conference on Big Data*, IEEE, 2017, P. 917–922

[DTC+18]   DANG, L. T. ; TONDL, M. ; CHIU, M. H. H. ; REVOTE, J. ; PATEN, B. ; TANO, V. ; TOKOLYI, A. ; BESSE, F. ; QUAIFE-RYAN, G. ; CUMMING, H. u. a.: TrawlerWeb: an online de novo motif discovery tool for next-generation sequencing datasets. In: *BMC genomics* 19 (2018), Nr. 1, P. 238. – Springer

[DTS+08]   DING, H. ; TRAJCEVSKI, G. ; SCHEUERMANN, P. ; WANG, X. ;
            KEOGH, E.: Querying and mining of time series data: experimental
            comparison of representations and distance measures. In: *Proceed-
            ings of the VLDB Endowment* 1 (2008), Nr. 2, P. 1542–1552. –
            VLDB Endowment

[DuA16]   DU, X. ; ANTHONY, B.: Controlled angular and radial scanning for
            super resolution concentric circular imaging. In: *Optics express* 24
            (2016), Nr. 20, P. 22581–22595

[DuC04]   DUZHIN, S. V. ; CHEBOTAREVSKIĬ, B. D.: *Transformation groups
            for beginners.* 25th ed. American Mathematical Soc., 2004. – ISBN
            978–0821836439

[EHD+02]   EADS, D. R. ; HILL, D. ; DAVIS, S. ; PERKINS, S. J. ; MA, J. ;
            PORTER, R. B. ; THEILER, J. P.: Genetic Algorithms and Sup-
            port Vector Machines for Time Series Classification, International
            Society for Optics and Photonics, 2002, P. 74–85

[ElB18]   ELISH, M. C. ; BOYD, D.: Situating methods in the magic of Big
            Data and AI. In: *Communication monographs* 85 (2018), Nr. 1, P.
            57–80. – Taylor & Francis

[ElE17]   ELLAHYANI, A. ; EL ANSARI, M.: Mean shift and log-polar trans-
            form for road sign detection. In: *Multimedia Tools and Applications*
            76 (2017), Nr. 22, P. 24495–24513. – Springer

[ElR83]   ELLIOTT, D. F. ; RAO, K. R.: *Fast transforms algorithms, analyses,
            applications.* Elsevier, 1983. – ISBN 0122370805

[EPN+15]   EN, S. ; PETITJEAN, C. ; NICOLAS, S. ; HEUTTE, L.: Segmentation-
            free pattern spotting in historical document images. In: *13th Inter-
            national Conference on Document Analysis and Recognition*, IEEE,
            2015, P. 606–610

[EsA12]   ESLING, P. ; AGON, C.: Time-series data mining, ACM, 2012, P.
            1–34

[FAS+06]   FERREIRA, P. G. ; AZEVEDO, P. J. ; SILVA, C. G. ; BRITO, R. M.
            M.: Mining approximate motifs in time series. In: *Proceedings of
            the 9th International Conference on Discovery Science.* Berlin and
            Heidelberg : Springer-Verlag, 2006 (DS'06), P. 89–101

[FaV17]   FAKHRAZARI, A. ; VAKILZADIAN, H.: A survey on time series data
            mining. In: *IEEE International Conference on Electro Information
            Technology*, IEEE, 2017, P. 476–481

[FeN19]   FENG, T. ; NARAYANAN, S. S.:   Discovering Optimal Variable-
               length Time Series Motifs in Large-scale Wearable Recordings of
               Human Bio-behavioral Signals. In: *IEEE International Conference
               on Acoustics, Speech and Signal Processing*, IEEE, 2019, P. 7615–
               7619

[FGN+09]   FUCHS, E. ; GRUBER, T. ; NITSCHKE, J. ; SICK, B.: On-line motif
               detection in time series with swiftmotif. In: *Pattern Recognition* 42
               (2009), Nr. 11, P. 3015–3031. – Elsevier

[Fis36]   FISHER, R. A.:   The use of multiple measurements in taxonomic
               problems. In: *Annals of eugenics* 7 (1936), Nr. 2, P. 179–188. –
               Wiley Online Library

[FJH+18]   FU, Q. ; JING, B. ; HE, P. ; SI, S. ; WANG, Y.:   Fault Feature
               Selection and Diagnosis of Rolling Bearings Based on EEMD and
               Optimized Elman_AdaBoost Algorithm. In: *IEEE Sensors Journal*
               18 (2018), Nr. 12, P. 5024–5034. – IEEE

[FKL+08]   FU, A. W.-C. ; KEOGH, E. ; LAU, L. Y. ; RATANAMAHATANA,
               C. A. ; WONG, R. C.-W.: Scaling and time warping in time series
               querying. In: *The International Journal on Very Large Data Bases*
               17 (2008), Nr. 4, P. 899–921. – Springer-Verlag New York, Inc.

[Fla01]   FLANDRIN, P.: Time frequency and chirps. In: *Wavelet Applications
               VIII* Bd. 4391 International Society for Optics and Photonics, 2001,
               P. 161–175. – SPIE

[FNR16]   FRANCIS, C. R. ; NAIR, V. V. ; RADHIKA, S.:   A scale invari-
               ant technique for detection of voice disorders using Modified Mellin
               Transform. In: *International Conference on Emerging Technological
               Trends*, IEEE, 2016, P. 1–6

[FoH10]   FOBER, T. ; HÜLLERMEIER, E.: Similarity measures for protein
               structures based on fuzzy histogram comparison. In: *International
               conference on fuzzy systems*, IEEE, 2010, P. 1–7

[Fou22]   FOURIER, J.:   *Theorie analytique de la chaleur, par M. Fourier.*
               Chez Firmin Didot, père et fils, 1822

[Fou78]   FOURIER, J.:   *The analytical theory of heat.* The University Press,
               1878. – ISBN 0486495310

[FrB70]   FRASER, A. ; BURNELL, D.: *Computer models in genetics.* Mcgraw-
               Hill Book Co., New York., 1970. – ISBN 978–0–07–021904–5

[Fri41]   FRINK, O.: Series Expansions in Linear Vector Space. In: *American
               Journal of Mathematics* 63 (1941), Nr. 1, P. 87–100. – Johns Hopkins
               University Press

[Fu11]   FU, T. C.: A review on time series data mining. In: *Engineering Applications of Artificial Intelligence* Bd. 24, Elsevier, 2011, P. 164–181

[FWDC+18]   FISCON, G. ; WEITSCHEK, E. ; DE COLA, M. C. ; FELICI, G. ; BERTOLAZZI, P.: An integrated approach based on EEG signals processing combined with supervised methods to classify Alzheimer's disease patients. In: *IEEE International Conference on Bioinformatics and Biomedicine*, IEEE, 2018, P. 2750–2752

[Gab09]   GABER, M. M.: *Scientific data mining and knowledge discovery.* 1st ed. Springer, 2009. – ISBN 978–3–642–42624–7

[Gal07]   GALUSHKIN, A. I.: *Neural networks theory.* 1st ed. Springer Science & Business Media, 2007. – ISBN 978–3–642–08006–7

[GaL18]   GAO, Y. ; LIN, J.: HIME: discovering variable-length motifs in large-scale time series. In: *Knowledge and Information Systems* (2018), P. 1–30. – Springer

[GaL19]   GAO, Y. ; LIN, J.: Discovering Subdimensional Motifs of Different Lengths in Large-Scale Multivariate Time Series. In: *IEEE International Conference on Data Mining*, IEEE, 2019, P. 220–229

[GaM56]   GARNER, W. R. ; MCGILL, W. J.: The relation between information and variance analyses. In: *Psychometrika* 21 (1956), Nr. 3, P. 219–228. – Springer

[GBZ18]   GAI, S. ; BAO, Z. ; ZHANG, K.: Vector extension of quaternion wavelet transform and its application to colour image denoising. In: *IET Signal Processing* 13 (2018), Nr. 2, P. 133–140. – IET

[GCK+18]   GALLAGHER, M. ; CHANDRA, S. ; KAPSALAS, Pe. ; HUGHES, C. ; GLAVIN, M. ; JONES, E.: Fourier Mellin transform characterisation in the automotive environment. In: *Signal, Image and Video Processing* 12 (2018), Nr. 8, P. 1587–1594. – Springer

[GGS+09]   GLOCK, S. ; GILLICH, E. ; SCHAEDE, J. ; LOHWEG, V.: Feature extraction algorithm for banknote textures based on incomplete shift invariant wavelet packet transform. In: *Proceedings 31st DAGM Symposium Pattern Recognition* Bd. 5748. Springer Berlin Heidelberg, 2009, P. 422–431

[GLR17]   GAO, Y. ; LIN, J. ; RANGWALA, H.: Iterative grammar-based framework for discovering variable-length time series motifs. In: *IEEE International Conference on Data Mining*, IEEE, 2017, P. 111–116

[GMVEFC16]   GUALSAQUÍ MIRANDA, M. V. ; VIZCAÍNO ESPINOSA, I. P. ; FLORES CALERO, M. J.: ECG signal features extraction. In: *IEEE Ecuador Technical Chapters Meeting* Bd. 1, IEEE, 2016, P. 1–6

[Gru05]   GRUNWALD, P.: A tutorial introduction to the minimum description length principle. In: *Advances in Minimum Description Length: Theory and Applications* (2005), P. 3–81

[GSI+14]   GULATI, S. ; SERRÀ, J. ; ISHWAR, V. ; SERRA, X.: Mining melodic patterns in large audio collections of indian art music. In: *Tenth International Conference on Signal-Image Technology and Internet-Based Systems*, IEEE, 2014, P. 264–271

[GSST16]   GRABOCKA, J. ; SCHILLING, N. ; SCHMIDT-THIEME, L.: Latent time-series motifs. In: *ACM Transactions on Knowledge Discovery from Data* 11 (2016), Nr. 1, P. 6. – ACM

[GSW+14]   GRABOCKA, J. ; SCHILLING, N. ; WISTUBA, M. ; SCHMIDT-THIEME, L.: Learning Time-series Shapelets. In: *Proceedings of the 20th ACM SIGKDD International Conference on Knowledge Discovery and Data Mining*, ACM, 2014 (KDD '14), P. 392–401

[HaB15]   HASSAN, A. R. ; BHUIYAN, M. I. H.: Dual tree complex wavelet transform for sleep state identification from single channel electroencephalogram. In: *IEEE International Conference on Telecommunications and Photonics*, IEEE, 2015, P. 1–5

[HaL16]   HARDY, G. H. ; LITTLEWOOD, J. E.: Contributions to the theory of the Riemann zeta-function and the theory of the distribution of primes. In: *Acta Mathematica* 41 (1916), P. 119–196. – Institut Mittag-Leffler

[HCZ+13]   HAO, Y. ; CHEN, Y. ; ZAKARIA, J. ; HU, B. ; RAKTHANMANON, T. ; KEOGH, E.: Towards never-ending learning from time series streams. In: *Proceedings of the 19th ACM SIGKDD international conference on Knowledge discovery and data mining*, ACM, 2013, P. 874–882

[HGD19]   HATAMI, N. ; GAVET, Y. ; DEBAYLE, J.: Bag of recurrence patterns representation for time-series classification. In: *Pattern Analysis and Applications* 22 (2019), Nr. 3, P. 877–887. – Springer

[HH19]   HUMEAU-HEURTIER, A.: Texture feature extraction methods: A survey. In: *IEEE Access* 7 (2019), P. 8975–9000. – IEEE

[Hil53]   HILBERT, D.: *Grundzüge einer allgemeinen Theorie der linearen Integralgleichungen, Chelsea Pub.* 1953

[HKV08]   HAVELOCK, D. ; KUWANO, S. ; VORLÄNDER, M.: *Handbook of signal processing in acoustics.* 2th ed. Springer Science & Business Media, 2008. – ISBN 978–0–387–30441–0

[HoA04]   HORST, B. ; ABRAHAM, K.: *Data mining in time series databases*.
             Bd. 57. World scientific, 2004. – ISBN 981–238–290–9

[HSYP+13]   HAO, Y. ; SHOKOOHI-YEKTA, M. ; PAPAGEORGIOU, G. ; KEOGH,
             E.: Parameter-free audio motif discovery in large data archives. In:
             IEEE 13th International Conference on Data Mining, IEEE, 2013,
             P. 261–270

[HTW15]   HE, H. ; TAN, Y. ; WANG, Y.: Optimal base wavelet selection for
             ECG noise reduction using a comprehensive entropy criterion. In:
             Entropy 17 (2015), Nr. 9, P. 6093–6109. – Multidisciplinary Digital
             Publishing Institute

[IBM20]   IBM: *IBM Corporation 1994-2020*.   https://www.ibm.com.
             Version: 2020. – Last access: 20.11.2020

[IMD+18]   IMANI, S. ; MADRID, F. ; DING, W. ; CROUTER, S. ; KEOGH, E.:
             Matrix Profile XIII: Time Series Snippets: A New Primitive for
             Time Series Data Mining. In: *IEEE International Conference on
             Big Knowledge*, IEEE, 2018, P. 382–389

[ImK19]   IMANI, S. ; KEOGH, E.: Matrix Profile XIX: Time Series Semantic
             Motifs: A New Primitive for Finding Higher-Level Structure in Time
             Series. In: *2019 IEEE International Conference on Data Mining
             (ICDM)*, IEEE, 2019, P. 329–338

[InH20]   INÈS SILVA, M. ; HENRIQUES, R.:   Exploring time-series motifs
             through DTW-SOM. In: *2020 International Joint Conference on
             Neural Networks (IJCNN)*, IEEE, 2020, P. 1–8

[Iva67]   IVAKHNENKO, Lapa V. A. G. G. A. G.: *Cybernetics and forecasting
             techniques*. American Elsevier Publishing Company, 1967 (Modern
             analytic and computational methods in science and mathematics).
             – ISBN 978–0444000200. – Translated by McDonough, R.N.

[Ize08]   IZENMAN, A. J.: Linear Discriminant Analysis. In: *Modern Multi-
             variate Statistical Techniques: Regression, Classification, and Man-
             ifold Learning*, Springer New York, 2008, P. 237–280

[Jac01]   JACCARD, Paul: Étude comparative de la distribution florale dans
             une portion des Alpes et des Jura. 37 (1901), P. 547–579. – Bull
             Soc Vaudoise Sci Nat

[JeC01]   JENSEN, A. ; COUR-HARBO, A. la: *Ripples in mathematics: the
             discrete wavelet transform*. Springer Verlag Berlin Heidelberg, 2001.
             – ISBN 3–540–41662–5

[Joy03]   JOYCE, J.: Bayes' Theorem. In: ZALTA, E. N. (Hrsg.): *The Stanford
             Encyclopedia of Philosophy*. Spring 2019. Metaphysics Research
             Lab, Stanford University, 2003

[JZP+19]   JI, C. ; ZHAO, C. ; PAN, L. ; LIU, S. ; YANG, C.i ; MENG, X.: A just-
             in-time shapelet selection service for online time series classification.
             In: *Computer Networks* 157 (2019), P. 89–98. – Elsevier

[KCH+04]   KEOGH, E. ; CHU, S. ; HART, D. ; PAZZANI, M.: Segmenting time
             series: A survey and novel approach. In: *Data mining in time series
             databases*, World Scientific, 2004, P. 1–21

[KCP+01]   KEOGH, E. ; CHAKRABARTI, K. ; PAZZANI, M. ; MEHROTRA, S.:
             Dimensionality reduction for fast similarity search in large time se-
             ries databases. In: *Knowledge and information Systems* 3 (2001),
             Nr. 3, P. 263–286. – Springer

[Keh11]   KEHTARNAVAZ, N.:   *Digital signal processing system design:
             LabVIEW-based hybrid programming.* 2nd ed. Elsevier, 2011. –
             ISBN 978–0–12–374490–6

[KeK03]   KEOGH, E. ; KASETTY, Sh.:   On the need for time series data
             mining benchmarks: a survey and empirical demonstration. In:
             Data Mining and Knowledge Discovery 7 (2003), Nr. 4, P. 349–371.
             – Springer

[Keo03]   KEOGH, E.: Efficiently finding arbitrarily scaled patterns in mas-
             sive time series databases. In: *Knowledge Discovery in Databases.*
             Springer Berlin Heidelberg, 2003, P. 253–265

[KeR05]   KEOGH, E. ; RATANAMAHATANA, C. A.: Exact indexing of dynamic
             time warping. In: *Knowledge and information Systems* 7 (2005), Nr.
             3, P. 358–386. – Springer

[KGK19]   KAMGAR, K. ; GHARGHABI, S. ; KEOGH, E.: Matrix Profile XV:
             Exploiting Time Series Consensus Motifs to Find Structure in Time
             Series Sets. In: *2019 IEEE International Conference on Data Min-
             ing (ICDM)*, IEEE, 2019, P. 1156–1161

[KhK11]   KHAMSI, M. A. ; KIRK, W. A.: *An introduction to metric spaces
             and fixed point theory.* Bd. 53. John Wiley & Sons, 2011. – ISBN
             0–471–41825–0

[Kin00]   KINGSBURY, N.: Complex wavelets and shift invariance. In: *IEEE
             Seminar on Time-scale and Time-Frequency Analysis and Applica-
             tions*, IEEE, 2000, P. 5–1

[Kin01]   KINGSBURY, N.: Complex wavelets for shift invariant analysis and
             filtering of signals. In: *Applied and computational harmonic analysis*
             10 (2001), Nr. 3, P. 234–253. – Elsevier

[Kin03]   KINGSBURY, N.: Design of q-shift complex wavelets for image pro-
               cessing using frequency domain energy minimization. In: *Proceed-
               ings International Conference on Image Processing* Bd. 1, IEEE,
               2003, P. I–1013

[KJF97]   KORN, F. ; JAGADISH, H. V. ; FALOUTSOS, C.: Efficiently support-
               ing ad hoc queries in large datasets of time sequences. In: *ACM
               SIGMOD Record* 26 (1997), Nr. 2, P. 289–300. – ACM

[KS+83]   KEMENY, J. G. ; SNELL, J. L. u. a.: *Finite markov chains.* Bd.
               3rd ed. Spinger-Verlag, 1983. – ISBN 0–387–90192–2. – Originally
               published in 1960 by Van Nostrand, Princeton, NJ.

[KST+14]   KERSCHEN, G. ; SHAW, S.W. ; TOUZÉ, C. ; GENDELMAN, O.V.
               ; COCHELIN, B. ; VAKAKIS, A.F.: *Modal analysis of nonlinear
               mechanical systems.* Bd. Series Vol. 555. Springer Verlag, 2014. –
               ISBN 978–3–7091–1791–0

[Kun79]   KUNZ, H. O.:   On the Equivalence Between One-Dimensional
               Discrete Walsh-Hadamard and Multidimensional Discrete Fourier
               Transforms. In: *IEEE Transactions on Computers* C-28 (1979), Nr.
               3, P. 267–268. – IEEE

[KuS12]   KUMAR, N. A. M. ; SATHIDEVI, P. S.: Image match using wavelet-
               colour SIFT features. In: *7th IEEE International Conference on
               Industrial and Information Systems*, IEEE, 2012, P. 1–6

[KWX+09]   KEOGH, E. ; WEI, L. ; XI, X. ; VLACHOS, M. ; LEE, S.-H. ;
               PROTOPAPAS, P.: Supporting exact indexing of arbitrarily rotated
               shapes and periodic time series under Euclidean and warping dis-
               tance measures. In: *The International Journal on Very Large Data
               Bases, The VLDB Journal.* 18 (2009), Nr. 3, P. 611–630. – Springer-
               Verlag New York, Inc.

[LAE+18]   LAGRANGE, M. ; ANDRIEU, H. ; EMMANUEL, I. ; BUSQUETS, G.
               ; LOUBRIÉ, S.:   Classification of rainfall radar images using the
               scattering transform. In: *Journal of Hydrology* 556 (2018), P. 972–
               979. – Elsevier

[LaS18]   LAWSON, K. K. K. ; SRINIVASAN, M. V.: A robust dual-axis vir-
               tual reality platform for closed-loop analysis of insect flight. In:
               *2018 IEEE International Conference on Robotics and Biomimetics
               (ROBIO)*, IEEE, 2018, P. 262–267

[LaW66]   LANCE, G. N. ; WILLIAMS, W. T.:   Computer programs for hi-
               erarchical polythetic classification ("similarity analyses"). In: *The
               Computer Journal* 9 (1966), Nr. 1, P. 60–64. – The British Com-
               puter Society

[LCG+16]   LI, N. ; CRANE, M. ; GURRIN, C. ; RUSKIN, H. J.: Finding Motifs
            in Large Personal Lifelogs. In: *Proceedings of the 7th Augmented
            Human International Conference*, ACM, 2016 (VAH '16), P. 9:1–9:8

[LCS11]   LEUTENEGGER, S. ; CHLI, M. ; SIEGWART, R. Y.: BRISK: Binary
            robust invariant scalable keypoints. In: *International conference on
            computer vision*, IEEE, 2011, P. 2548–2555

[LDM04]   LOHWEG, V. ; DIEDERICHS, C. ; MÜLLER, D.:   Algorithms for
            hardware-based pattern recognition. In: *EURASIP Journal on Ap-
            plied Signal Processing* (2004), P. 1912–1920. – Hindawi Publishing
            Corp.

[Lev66]   LEVENSHTEIN, V. I.: Binary codes capable of correcting deletions,
            insertions and reversals. In: *Soviet physics doklady* Bd. 10, 1966, P.
            707

[LFW+19]   LUU, V.-T. ; FORESTIER, G. ; WEBER, J. ; BOURGEOIS, P. ; DJELIL,
            F. ; MULLER, P.-A.: A review of alignment based similarity mea-
            sures for web usage mining. In: *Artificial Intelligence Review* (2019),
            P. 1–23. – Springer

[LiB15]   LINES, J. ; BAGNALL, A.:   Time series classification with ensem-
            bles of elastic distance measures. In: *Data Mining and Knowledge
            Discovery* 29 (2015), Nr. 3, P. 565–592. – Springer

[LiD12]   LI, C.-R. ; DENG, Y.-H: Rotation-invariant texture Image Clas-
            sification using R-transform. In: *2nd International Conference on
            Uncertainty Reasoning and Knowledge Engineering*, IEEE, 2012, P.
            271–274

[LiL10]   LIN, J. ; LI, Y.: Finding approximate frequent patterns in stream-
            ing medical data. In: *IEEE 23rd International Symposium on
            Computer-Based Medical Systems*, IEEE, 2010, P. 13–18

[LiL19]   LI, X. ; LIN, J.: Linear Time Motif Discovery in Time Series. In:
            *Proceedings of the 2019 SIAM International Conference on Data
            Mining*, SIAM, 2019, P. 136–144

[LKL+03]   LIN, J. ; KEOGH, E. ; LONARDI, S. ; CHIU, B.: A symbolic repre-
            sentation of time series, with implications for streaming algorithms.
            In: *Proceedings of the 8th ACM SIGMOD Workshop on Research
            Issues in Data Mining and Knowledge Discovery*, ACM, 2003, P.
            2–11

[LKW+07]   LIN, J. ; KEOGH, E. ; WEI, L. ; LONARDI, S.: Experiencing SAX: a
            novel symbolic representation of time series. In: *Data Mining and
            Knowledge Discovery* 15 (2007), Nr. 2, P. 107–144. – Springer

[LLC+15]   LIU, B. ; LI, J. ; CHEN, C. ; TAN, W. ; CHEN, Q. ; ZHOU, M.:
            Efficient Motif Discovery for large-scale time series in healthcare.
            In: *IEEE Transactions on Industrial Informatics* 11 (2015), Nr. 3,
            P. 583–590. – IEEE

[LLG+14]   LI, C. ; LI, J. ; GAO, D. ; FU, B.:   Rapid-transform based rota-
            tion invariant descriptor for texture classification under non-ideal
            conditions. In: *Pattern Recognition* 47 (2014), Nr. 1, P. 313–325. –
            Elsevier

[LLO12]   LI, Y. ; LIN, J. ; OATES, T.: Visualizing variable-length time series
            motifs. In: *SIAM International Conference on Data Mining*, SIAM,
            2012, P. 895–906

[LLR+17]   LU, n. ; LI, T. ; REN, X. ; MIAO, H.:   A Deep Learning Scheme
            for Motor Imagery Classification based on Restricted Boltzmann
            Machines. In: *IEEE Transactions on Neural Systems and Rehabili-
            tation Engineering* 25 (2017), Nr. 6, P. 566–576. – IEEE

[LLW08]   LIU, X. ; LIN, X. ; WANG, H.:   Novel Online Methods for Time
            Series Segmentation.   In: *IEEE Transactions on Knowledge and
            Data Engineering* 20 (2008), Nr. 12, P. 1616–1626. – IEEE

[LLZ+02]   LI, T. ; LI, Q. ; ZHU, S. ; OGIHARA, M.: A survey on wavelet appli-
            cations in data mining. In: *ACM SIGKDD Explorations Newsletter*
            4 (2002), Nr. 2, P. 49–68. – ACM

[LoM02]   LOHWEG, V. ; MÜLLER, D.: A complete set of translation invariants
            based on the cyclic correlation property of the generalised circular
            transforms. In: *Proceeding 6th Digital Image Computing Techniques
            and Applications (DICTA'02)* (2002), P. 134–138

[Low99]   LOWE, D.G: Object recognition from local scale-invariant features.
            In: *The proceedings of the seventh IEEE international conference
            on Computer vision* Bd. 2, IEEE, 1999, P. 1150–1157

[LVK+04]   LIN, J. ; VLACHOS, M. ; KEOGH, E. ; GUNOPULOS, D.: Iterative
            incremental clustering of time series.   In: *Advances in Database
            Technology-EDBT*, Springer Berlin Heidelberg, 2004, P. 106–122

[LYL+05]   LIU, Z. ; YU, J. X. ; LIN, X. ; LU, H. ; WANG, W.: Locating motifs
            in time-series data. In: *Advances in Knowledge Discovery and Data
            Mining*, Springer Berlin Heidelberg, 2005 (3518), P. 343–353

[LZP+18]   LINARDI, M. ; ZHU, Y. ; PALPANAS, T. ; KEOGH, E.:   Matrix
            profile X: VALMOD-scalable discovery of variable-length motifs in
            data series. In: *Proceedings of the 2018 International Conference
            on Management of Data*, ACM, 2018, P. 1053–1066

[Mad97]   MADISETTI, V.: *The digital signal processing handbook.* CRC press, 1997. – ISBN 0849385725

[Mai14]   MAIER, A.: Online passive learning of timed automata for cyber-physical production systems. In: *12th IEEE International Conference on Industrial Informatics*, IEEE, 2014, P. 60–66

[Mal89]   MALLAT, S.: Multiresolution approximations and wavelet orthonormal bases of L^2 (R). In: *Transactions of the American mathematical society* 315 (1989), Nr. 1, P. 69–87

[Mal08]   MALLAT, S.: *A wavelet tour of signal processing: the sparse way.* 3rd, ed. Academic press, 2008. – ISBN 978–0–12–374370–1

[Mal12]   MALLAT, S.: Group invariant scattering. In: *Communications on Pure and Applied Mathematics* 65 (2012), Nr. 10, P. 1331–1398. – Wiley Online Library

[MaR10]   MAIMON, O. ; ROKACH, L.: *Data mining and knowledge discovery handbook.* 2nd ed. 2010

[MaW15]   MARZ, N. ; WARREN, J.: *Big Data: principles and best practices of scalable realtime data systems.* 1st ed. Manning Publications Co., 2015. – ISBN 9781617290343

[MBM09]   MÜLLER, F. ; BELILOVSKY, E. ; MERTINS, A.: Generalised cyclic transformations in speaker-independent speech recognition. In: *IEEE Workshop on Automatic Speech Recognition Understanding*, IEEE, 2009, P. 211–215

[MCAER^{+}18] MUEEN, A. ; CHAVOSHI, N. ; ABU-EL-RUB, N. ; HAMOONI, H. ; MINNICH, A. ; MACCARTHY, J.: Speeding up dynamic time warping distance for sparse time series data. In: *Knowledge and Information Systems* 54 (2018), Nr. 1, P. 237–263. – Springer

[Mck19]   MCKAY, C.: *Probability and Statistics.* 1st ed. ED - Tech Press, 2019. – ISBN 978–1–83947–330 2

[Mey93]   MEYER, Y.: *Wavelets-algorithms and applications.* Society for Industrial & Applied Mathematics, 1993 (Miscellaneous Bks). – ISBN 978–0898713091

[MIE^{+}07a] MINNEN, D. ; ISBELL, C. ; ESSA, I. ; STARNER, T.: Detecting subdimensional motifs: An efficient algorithm for generalised multivariate pattern discovery. In: *Seventh IEEE International Conference on Data Mining*, IEEE, 2007, P. 601–606

[MIE⁺07b]   MINNEN, D. ; ISBELL, C. ; ESSA, I. ; STARNER, T.:   Discovering
             multivariate motifs using subsequence density estimation and greedy
             mixture learning.   In: *Proceedings of the National Conference on
             Artificial Intelligence* Bd. 22, AAAI Press, 2007, P. 615–620

[MIM⁺19]   MADRID, F. ; IMANI, S. ; MERCER, R. ; ZIMMERMAN, Z. ; SHAK-
             IBAY, N. ; KEOGH, E.: Matrix Profile XX: Finding and Visualizing
             Time Series Motifs of All Lengths using the Matrix Profile. In: *2019
             IEEE International Conference on Big Knowledge (ICBK)*, IEEE,
             2019, P. 175–182

[Mit98]   MITCHELL, M.: *An introduction to genetic algorithms.* MIT press,
             1998 (A Bradford book). – ISBN 9780262631853

[MJ51]   MASSEY JR, F. J.:   The Kolmogorov-Smirnov test for goodness of
             fit. In: *Journal of the American statistical Association* 46 (1951),
             Nr. 253, P. 68–78. – Taylor & Francis Group

[MKBS09]   MUEEN, A. ; KEOGH, E. ; BIGDELY-SHAMLO, N.:   Finding time
             series motifs in disk-resident data. In: *9th IEEE International Con-
             ference on Data Mining*, IEEE, 2009, P. 367–376

[MKZ⁺09]   MUEEN, A. ; KEOGH, E. ; ZHU, Q. ; CASH, S. ; WESTOVER, M. B.:
             Exact discovery of time series motifs. In: *Proceedings of the SIAM
             International Conference on Data Mining*, SIAM, 2009, P. 473–484

[MoN14]   MOHAMMAD, Y. ; NISHIDA, T.: Exact discovery of length-range mo-
             tifs. In: *Asian Conference on Intelligent Information and Database
             Systems.* Springer International Publishing, 2014, P. 23–32

[MoN16]   MOHAMMAD, Y. ; NISHIDA, T.: Exact multi-length scale and mean
             invariant motif discovery. In: *Applied Intelligence* 44 (2016), Nr. 2,
             P. 322–339. – Springer

[MSE⁺06]   MINNEN, D. ; STARNER, T. ; ESSA, I. ; ISBELL, C.:   Discovering
             characteristic actions from on-body sensor data.   In: *Proceedings
             OF IEEE International Symposium On Wearable Computing*, IEEE,
             2006, P. 11–18

[MuC15]   MUEEN, A. ; CHAVOSHI, N.: Enumeration of time series motifs of
             all lengths. In: *Knowledge and information Systems* 45 (2015), Nr.
             1, P. 105–132. – Springer

[Mue14]   MUEEN, A.: Time series motif discovery: dimensions and applica-
             tions. In: *Wiley Interdisciplinary Reviews: Data Mining and Knowl-
             edge Discovery* 4 (2014), Nr. 2, P. 152–159. – ISSN 1942–4795. –
             Wiley Online Library

[MuK10]   MUEEN, A. ; KEOGH, E.: Online discovery and maintenance of time series motifs. In: *Proceedings of the 16th International Conference on Knowledge Discovery and Data Mining*, ACM, 2010 (KDD '10), P. 1089–1098

[Mül07]   MÜLLER, M.: Dynamic time warping. In: *Information retrieval for music and motion* (2007), P. 69–84. – Springer

[Mül13]   MÜLLER, F.: *Invariant features and enhanced speaker normalisation for automatic speech recognilion*. Logos Verlag Berlin GmbH, 2013. – ISBN 3832533192

[MüM11]   MÜLLER, F. ; MERTINS, A.: Robust continuous speech recognition through combination of invariant-feature based systems. In: *Proc. German Conf. Speech Signal Processing (ESSV 2011)*, TUDPress, 2011, P. 229–236

[MVN+11]   MAIER, A. ; VODENCAREVIC, A. ; NIGGEMANN, O. ; JUST, R. ; JAEGER, M.: Anomaly detection in production plants using timed automata. In: *8th International Conference on Informatics in Control, Automation and Robotics*, 2011, P. 363–369

[MZY+17]   MUEEN, A. ; ZHU, Y. ; YEH, M. ; KAMGAR, K. ; VISWANATHAN, K. ; GUPTA, C. ; KEOGH, E.: *The Fastest Similarity Search Algorithm for Time Series Subsequences under Euclidean Distance*. 2017. – http://www.cs.unm.edu/~mueen/FastestSimilaritySearch.html (Last access:19.01.2020)

[NNR11]   NUNTHANID, P. ; NIENNATTRAKUL, V. ; RATANAMAHATANA, C. A.: Discovery of variable length time series motif. In: *8th International Conference on Electrical Engineering/Electronics, Computer, Telecommunications and Information Technology*, IEEE, 2011, P. 472–475

[NNR12]   NUNTHANID, P. ; NIENNATTRAKUL, V. ; RATANAMAHATANA, C. A.: Parameter-free motif discovery for time series data. In: *9th International Conference on Electrical Engineering/Electronics, Computer, Telecommunications and Information Technology*, IEEE, 2012, P. 1–4

[NRG13]   NITHYA, A. ; RANJANI, J. J. ; GOWTHAMI, D.: Survey on rotation, scaling, and translation invariant watermarking algorithm. In: *Image* 2 (2013), Nr. 11

[NSV+12]   NIGGEMANN, O. ; STEIN, B. ; VODENCAREVIC, A. ; MAIER, Al. ; KLEINE BÜNING, H.: Learning Behavior Models for Hybrid Timed Systems. In: *Twenty-Sixth Conference on Artificial Intelligence AAAI* Bd. 2, 2012, P. 1083–1090

[Oat02]   OATES, T.: PERUSE: An unsupervised algorithm for finding recur-
               ring patterns in time series. In: *IEEE International Conference on
               Data Mining*, IEEE, 2002, P. 330–337

[OpL81]   OPPENHEIM, A. V. ; LIM, J. S.: The importance of phase in signals.
               In: *Proceedings of the IEEE* 69 (1981), Nr. 5, P. 529–541. – IEEE

[OpS89]   OPPENHEIM, A. V. ; SCHAFER, R. W.: *Discrete-time signal pro-
               cessing.* 2nd ed. Prentice-Hall, 1989 (Prentice-Hall signal processing
               series). – ISBN 978–01317549207

[OZH⁺19]   OYALLON, E. ; ZAGORUYKO, S. ; HUANG, G. ; KOMODAKIS, N. ;
               LACOSTE-JULIEN, S. ; BLASCHKO, M. ; BELILOVSKY, E.: Scattering
               Networks for Hybrid Representation Learning. In: *IEEE Transac-
               tions on Pattern Analysis and Machine Intelligence* 41 (2019), Nr.
               9, P. 2208–2221. – IEEE

[PaD94]   PATERSON, M. ; DANCIK, V.: *Longest common subsequences.* Bd.
               841. Springer Berlin Heidelberg, 1994. – ISBN 978–3–540–48663–3

[Par06]   PARSEVAL, M.-A.: Mèmoire sur les sèries et sur l'intègration
               complète d'une èquation aux diffèrences partielles linèaires du sec-
               ond ordre, à coefficients constants. In: *Mèm. près. par divers sa-
               vants, Acad. des Sciences, Paris,(1)* 1 (1806), P. 638–648

[PaS17]   PATEL, T. ; SHAH, B.: A survey on facial feature extraction tech-
               niques for automatic face annotation. In: *International Conference
               on Innovative Mechanisms for Industry Applications*, IEEE, 2017,
               P. 224–228

[PBD17]   PÉREZ, D. S. ; BROMBERG, F. ; DIAZ, C. A.: Image classifica-
               tion for detection of winter grapevine buds in natural conditions
               using scale-invariant features transform, bag of features and sup-
               port vector machines. In: *Computers and electronics in agriculture*
               135 (2017), P. 81–95. – Elsevier

[PCB18]   PHAN, T. ; CAILLAULT, É. P. ; BIGAND, A.: Comparative Study
               on Univariate Forecasting Methods for Meteorological Time Series.
               In: *26th European Signal Processing Conference*, IEEE, 2018, P.
               2380–2384

[Pea01]   PEARSON, K.: LIII. On lines and planes of closest fit to systems of
               points in space. In: *The London, Edinburgh, and Dublin Philosoph-
               ical Magazine and Journal of Science* 2 (1901), Nr. 11, P. 559–572.
               – Taylor & Francis

[Pea05]   PEARSON, K.: Skew variation, a rejoinder. In: *Biometrika* 4 (1905),
               Nr. 1-2, P. 169–212

[PfL18]   PFEIFER, A. ; LOHWEG, V.: Identifying Characteristic Gait Pat-
             terns in Real-World Scenarios. In: *PROCEEDINGS 28. WORK-
             SHOP COMPUTATIONAL INTELLIGENCE*, KIT Scientific Pub-
             lishing, 2018, P. 279–295

[PKL⁺02]   PATEL, P. ; KEOGH, E. ; LIN, J. ; LONARDI, S.: Mining motifs in
             massive time series databases. In: *Proceedings IEEE International
             Conference on Data Mining*, IEEE, 2002, P. 370–377

[PNA19]   PHIEN, N. N. ; NHAN, N. T. ; ANH, D. T.: An Efficient Method for
             Discovering Variable-Length Motifs in Time Series Based on Suffix
             Array, ACM, 2019 (SoICT 2019), P. 125–131

[PNP⁺12]   PHINYOMARK, A. ; NUIDOD, A. ; PHUKPATTARANONT, P. ; LIM-
             SAKUL, C.: Feature extraction and reduction of wavelet transform
             coefficients for EMG pattern classification. In: *Elektronika ir Elek-
             trotechnika* 122 (2012), Nr. 6, P. 27–32

[PoM07]   POLYANIN, A. D. ; MANZHIROV, A. V.: *Handbook of mathematics
             for engineers and scienctists.* Boca Raton : Chapman & Hall/CRC,
             2007. – ISBN 9781584885023

[Pou10]   POULARIKAS, A. D.: *Transforms and applications handbook.* CRC
             Press, 2010. – ISBN 9781420066524

[Pow11]   POWERS, D.: Evaluation: From Precision, Recall and F-Factor
             to ROC, Informedness, Markedness & Correlation. In: *Journal of
             Machine Learning Technologies* 2 (2011), P. 37–63

[RaH20]   RAHNAMA, J. ; HÜLLERMEIER, E.: Learning Tversky Similar-
             ity. In: *Information Processing and Management of Uncertainty in
             Knowledge-Based Systems*, Springer International Publishing, 2020.
             – ISBN 978–3–030–50143–3, P. 269–280

[RaK04]   RATANAMAHATANA, C. ; KEOGH, E.: Making time-series classifi-
             cation more accurate using learned constraints. In: *Proceedings of
             SIAM International Conference on Data Mining*, SIAM, 2004, P.
             11–22

[RaM07]   RATH, T. M. ; MANMATHA, R.: Word spotting for historical docu-
             ments. In: *International Journal on Document Analysis and Recog-
             nition* 9 (2007), Nr. 2, P. 139–152. – Springer

[RCL⁺20]   RONG, C. ; CHEN, Z. ; LIN, C. ; WANG, J.: Motif Discovery
             Using Similarity-Constraints Deep Neural Networks. In: *Interna-
             tional Conference on Database Systems for Advanced Applications*
             Springer, 2020, P. 587–603

[ReB69]   REITBOECK, H. ; BRODY, T.P.: A transformation with invariance under cyclic permutation for applications in pattern recognition. In: *Information and Control* 15 (1969), Nr. 2, P. 130–154. – Elsevier

[REJ+17]   RIAZ, S. ; ELAHI, H. ; JAVAID, K. ; SHAHZAD, T.: Vibration feature extraction and analysis for fault diagnosis of rotating machinery-a literature survey. In: *Asia Pacific Journal of Multidisciplinary Research* 5 (2017), Nr. 1, P. 103–110

[ReL15]   REN, H. ; LI, Z.-N.: Object tracking using structure-aware binary features. In: *IEEE International Conference on Multimedia and Expo*, IEEE, 2015, P. 1–6

[RGC16]   ROYER, A. ; GRAVIER, G. ; CLAVEAU, V.: Audio word similarity for clustering with zero resources based on iterative HMM classification. In: *IEEE International Conference on Acoustics, Speech and Signal Processing*, IEEE, 2016, P. 5340–5344

[RJV18]   REZAEE, M. J. ; JOZMALEKI, M. ; VALIPOUR, M.: Integrating dynamic fuzzy C-means, data envelopment analysis and artificial neural network to online prediction performance of companies in stock exchange. In: *Physica A: Statistical Mechanics and its Applications* 489 (2018), P. 78–93. – Elsevier

[RMB16]   RAY, P. ; MAITRA, A. K. ; BASURAY, A.: Entropy-based wavelet de-noising for partial discharge measurement application. In: *First International Conference on Control, Measurement and Instrumentation*, IEEE, 2016, P. 264–268

[RVR+20]   ROMERO, J. C. ; VILCHES, A. ; RODRÍGUEZ, A. ; NAVARRO, A. ; ASENJO, R.: ScrimpCo: scalable matrix profile on commodity heterogeneous processors. In: *The Journal of Supercomputing* (2020), P. 1–22. – Springer

[RZK11]   RAKTHANMANON, T. ; ZHU, Q. ; KEOGH, E.: Mining historical documents for near-duplicate figures. In: *2011 IEEE 11th International Conference on Data Mining*, IEEE, 2011, P. 557–566

[SaB18]   SATO, Y. ; BAO, Y.: High Speed and High Precision Facial Recognition Using Log-Polar Transformation. In: *International Conference on Sensor Networks and Signal Processing*, IEEE, 2018, P. 375–381

[SaC07]   SALVADOR, S. ; CHAN, P.: Toward accurate dynamic time warping in linear time and space. In: *Intelligent Data Analysis* 11 (2007), Nr. 5, P. 561–580. – IOS Press

[SBK05]   SELESNICK, I. W. ; BARANIUK, R. G. ; KINGSBURY, N. G.: The dual-tree complex wavelet transform. In: *IEEE Signal Processing Magazine* 22 (2005), Nr. 6, P. 123–151. – IEEE

[ScG16]   SCHLEGEL, D. ; GRISETTI, G.: Visual localization and loop closing
            using decision trees and binary features. In: *IEEE/RSJ Interna-
            tional Conference on Intelligent Robots and Systems*, IEEE, 2016,
            P. 4616–4623

[Sch90]   SCHWARZ, H. A.: Über ein die Flächen kleinsten Flächeninhalts be-
            treffendes Problem der Variationsrechnung. In: *Gesammelte Math-
            ematische Abhandlungen*. Springer, 1890, P. 223–269

[ScH12]   SCHÄFER, P. ; HÖGQVIST, M.: SFA: a symbolic fourier approxima-
            tion and index for similarity search in high dimensional datasets.
            In: *Proceedings of the 15th International Conference on Extending
            Database Technology*, ACM, 2012, P. 516–527

[Sch16]   SCHÄFER, P.: Scalable time series classification. In: *Data Mining
            and Knowledge Discovery* 30 (2016), Nr. 5, P. 1273–1298. – Springer

[SÇŞ15]   SEVINDIR, H. K. ; ÇETUNKAYA, S. ; ŞAYLI, Ö.: Wavelet transform
            based noise removal from ECG signal for accurate heart rate de-
            tection using ECG. In: *Medical Technologies National Conference*,
            IEEE, 2015, P. 1–4

[SeA16]   SERRÀ, J. ; ARCOS, J. L.: Particle swarm optimization for time
            series motif discovery. In: *Knowledge-Based Systems* 92 (2016), P.
            127–137. – Elsevier

[Sel01]   SELESNICK, I. W.: Hilbert transform pairs of wavelet bases. In:
            IEEE Signal Processing Letters 8 (2001), Nr. 6, P. 170–173

[SeR17]   SEETHARAMAN, P. ; RAFII, Z.: Cover song identification with 2D
            Fourier Transform sequences. In: *IEEE International Conference
            on Acoustics, Speech and Signal Processing*, IEEE, 2017, P. 616–620

[SGO20]   SEZER, O. B. ; GUDELEK, M. U. ; OZBAYOGLU, A. M.: Financial
            time series forecasting with deep learning: A systematic literature
            review: 2005–2019. In: *Applied Soft Computing* 90 (2020), P. 106181

[Sha01]   SHANNON, C. E.: A mathematical theory of communication. In:
            *Mobile Computing and Communications Review, Reprinted for the
            Bell System Technical Journal 1948* 5 (2001), Nr. 1, P. 3–55. – ACM

[ShK08]   SHIEH, J. ; KEOGH, E.: iSAX: indexing and mining terabyte sized
            time series. In: *Proceedings of the 14th ACM international con-
            ference on Knowledge discovery and data mining*, ACM, 2008, P.
            623–631

[SlB16]   SILVA, D. F. ; BATISTA, G. E. A. P. A.: Speeding up all-pairwise
            dynamic time warping matrix calculation. In: *Proceedings of the
            SIAM International Conference on Data Mining*, SIAM, 2016, P.
            837–845

[SiB18]   SILVA, D. F. ; BATISTA, G. E. A. P. A.: Elastic Time Series Motifs
           and Discords. In: *17th IEEE International Conference on Machine
           Learning and Applications*, 2018, P. 237–242. – IEEE

[SiD18]   SINGH, B. ; DAVIS, L. S.: An Analysis of Scale Invariance in Object
           Detection Â SNIP. In: *IEEE Conference on Computer Vision and
           Pattern Recognition*, IEEE, 2018, P. 3578–3587

[Sig02]   SIGGELKOW, S.: *Feature histograms for content-based image re-
           trieval*, University of Freiburg, Germany, Diss., 2002

[SiK17]   SINGH, A. ; KINGSBURY, N.: Dual-tree wavelet scattering net-
           work with parametric log transformation for object classification.
           In: *IEEE International Conference on Acoustics, Speech and Signal
           Processing*, IEEE, 2017, P. 2622–2626

[SiR15]   SIVARAKS, H. ; RATANAMAHATANA, C. A.: Robust and accurate
           anomaly detection in ECG artifacts using time series motif discov-
           ery. In: *Computational and mathematical methods in medicine* 2015
           (2015). – Hindawi Publishing Corporation

[SKA18]   SHARMA, P. ; KAPOOR, G. ; ALI, S.: Fault Detection on Series
           Capacitor Compensated Transmission Line using Walsh Hadamard
           Transform. In: *International Conference on Computing, Power and
           Communication Technologies*, IEEE, 2018, P. 754–759

[SKR17]   SILVEIRA, T. L. T. ; KOZAKEVICIUS, A. J. ; RODRIGUES, C. R.:
           Single-channel EEG sleep stage classification based on a streamlined
           set of statistical features in wavelet domain. In: *Medical & biological
           engineering & computing* 55 (2017), Nr. 2, P. 343–352. – Springer

[SLK18]   SONG, J. ; LIM, S. ; KIM, S.-W.: A Novel Join Technique for Similar-
           trend Searches Supporting Normalisation on Time-series Databases.
           In: *Proceedings of the 33rd Annual ACM Symposium on Applied
           Computing*, ACM, 2018 (SAC '18), P. 481–486

[SLW+14]   SENIN, P. ; LIN, J. ; WANG, X. ; OATES, T. ; GANDHI, S. ; BOEDI-
           HARDJO, A. P. ; CHEN, C. ; FRANKENSTEIN, S. ; LERNER, M.:
           Grammarviz 2.0: a tool for grammar-based pattern discovery in
           time series. In: *Joint European conference on machine learning and
           knowledge discovery in databases*, Springer, 2014, P. 468–472

[SMRA14]   SILVA, P. F. B. ; MARCAL, A. R. S. ; R. A. daSilva: *UCI Machine
           Learning Repository: Leaf Data Set*. https://archive.ics.uci.
           edu/ml/datasets/Leaf. Version: 2014. – Last access: 20.01.2020

[SMY+19]   SHAO, S. ; MCALEER, S. ; YAN, R. ; BALDI, P.: Highly Accurate
           Machine Fault Diagnosis Using Deep Transfer Learning. In: *IEEE*

Transactions on Industrial Informatics 15 (2019), Nr. 4, P. 2446–2455. – IEEE

[Spi15]   SPIEGEL, S.: *Time series distance measures.* 2015. – PhD Thesis, Technische Universität Berlin, Fakultät IV - Elektrotechnik und Informatik

[SrS18]   SRIVASTVA, R. ; SINGH, Y. N.: ECG biometric analysis using walsh–hadamard transform. In: *Advances in Data and Information Sciences.* Springer, 2018, P. 201–210

[StW09]   STAFFORD, C.A. ; WALKER, G.P.: Characterization and correlation of DC electrical penetration graph waveforms with feeding behavior of beet leafhopper, Circulifer tenellus. In: *Entomologia Experimentalis et Applicata* 130 (2009), Nr. 2, P. 113–129. – Wiley Online Library

[Sut97]   SUTER, B. W.: *Multirate and wavelet signal processing.* Elsevier, 1997. – ISBN 0–12–677560–6

[SYCC+15] SHOKOOHI-YEKTA, M. ; CHEN, Y. ; CAMPANA, B. ; HU, B. ; ZAKARIA, J. ; KEOGH, E.: Discovery of meaningful rules in time series. In: *Proceedings of the 21th ACM SIGKDD International Conference on Knowledge Discovery and Data Mining.* USA : ACM, 2015 (KDD '15), P. 1085–1094

[SYS+15]   SHUKLA, S. ; YADAV, R. N. ; SHARMA, J. ; KHARE, S.: Analysis of statistical features for fault detection in ball bearing. In: *IEEE International Conference on Computational Intelligence and Computing Research,* IEEE, 2015, P. 1–7

[TaL08]   TANG, H. ; LIAO, S. S.: Discovering original motifs with different lengths from time series. In: *Knowledge-Based Systems* 21 (2008), Nr. 7, P. 666–671. – Elsevier

[TDA15]   THOMAS, M. ; DAS, M. K. ; ARI, S.: Automatic ECG arrhythmia classification using dual tree complex wavelet based features. In: *AEU-International Journal of Electronics and Communications* 69 (2015), Nr. 4, P. 715–721. – Elsevier

[TDDL16]   TORKAMANI (DEPPE), S. ; DICKS, A. ; LOHWEG, V.: Anomaly detection on ATMs via time series motif discovery. In: *21th IEEE International Conference on Emerging Technologies and Factory Automation.* Berlin, 2016. – IEEE

[Tho65]   THOMPSON, S. P.: *Calculus made easy.* 3rd ed. Macmillan Education, 1965. – 185 P. – ISBN 978–1–349–00487–4

[TIU05]   TANAKA, Y. ; IWAMOTO, K. ; UEHARA, K.: Discovery of time-
            series motif from multi-dimensional data based on MDL principle.
            In: *Machine Learning* 58 (2005), Nr. 2-3, P. 269–300. – Springer

[TKP06]   TAY, D. B. H. ; KINGSBURY, N. ; PALANISWAMI, M.: Orthonor-
            mal Hilbert-Pair of Wavelets With (Almost) Maximum Vanishing
            Moments. In: *IEEE Signal Processing Letters* 13 (2006), Nr. 9, P.
            533–536. – IEEE

[ToL14]   TORKAMANI (DEPPE), S. ; LOHWEG, V.: Identification of multi-
            scale motifs. In: *24. Workshop Computational Intelligence* Bd. 50,
            KIT Scientific Publishing, 2014 (Schriftenreihe des Instituts für
            Angewandte Informatik - Automatisierungstechnik am Karlsruher
            Institut für Technologie), P. 277–297

[ToL15]   TORKAMANI (DEPPE), S. ; LOHWEG, V.: Shift-invariant fea-
            ture extraction for time-series motif discovery. In: *25. Workshop
            Computational Intelligence* Bd. 54, KIT Scientific Publishing, 2015
            (Schriftenreihe des Instituts für Angewandte Informatik - Automa-
            tisierungstechnik am Karlsruher Institut für Technologie), P. 23–45

[ToL17a]   TORKAMANI (DEPPE), S. ; LOHWEG, V.: Shift-Invariant Motif
            Discovery in Image Processing "Best Paper Award". In: *PESARO
            2017 The Seventh International Conference on Performance, Safety
            and Robustness in Complex Systems and Applications*, 2017

[ToL17b]   TORKAMANI (DEPPE), S. ; LOHWEG, V.: Survey on time series
            motif discovery. 7 (2017), Nr. 2, P. e1199. – Wiley Interdisciplinary
            Reviews: Data Mining and Knowledge Discovery

[ToL18]   TORKAMANI (DEPPE), S. ; LOHWEG, V.: Evaluation of Similarity
            Measures for Shift-Invariant Image Motif Discovery. In: *Interna-
            tional Journal On Advances in Intelligent Systems* Bd. 10, IARIA
            Journals, 2018, P. 434–446

[TrA17]   TRUONG, C. D. ; ANH, D. T.: Discovering Motifs in a Database
            of Shapes Under Rotation Invariant Dynamic Time Warping. In:
            *Proceedings of the Eighth International Symposium on Information
            and Communication Technology*, ACM, 2017 (SoICT 2017), P. 18–
            25

[Tur37]   TURING, A. M.: On computable numbers, with an application to the
            Entscheidungsproblem. In: *Proceedings of the London mathematical
            society* 2 (1937), Nr. 1, P. 230–265. – Wiley Online Library

[Tur50]   TURING, A. M.: The word problem in semi-groups with cancella-
            tion. In: *Annals of Mathematics* (1950), P. 491–505

[Tya17]   TYAGI, V.: *Content-Based Image Retrieval*. 1st ed. Springer Singapore, 2017. – ISBN 978–981–10–6759–4

[Uhl18]   UHLMANN, J.: A generalised matrix inverse that is consistent with respect to diagonal transformations. In: *SIAM Journal on Matrix Analysis and Applications* 39 (2018), Nr. 2, P. 781–800. – SIAM

[UmY18]   UMAM, A. K. ; YUNUS, M.: Quaternion wavelet transform for image denoising. In: *Journal of Physics: Conference Series* Bd. 974, IOP Publishing, 2018

[Vai90]   VAIDYANATHAN, P. P.:   Multirate digital filters, filter banks, polyphase networks, and applications: a tutorial. In: *Proceedings of the IEEE* 78 (1990), Nr. 1, P. 56–93. – IEEE

[VaP17]   VAN NOORD, N. ; POSTMA, E.: Learning scale-variant and scale-invariant features for deep image classification. In: *Pattern Recognition* 61 (2017), P. 583–592. – Elsevier

[VAS09]   VAHDATPOUR, A. ; AMINI, N. ; SARRAFZADEH, M.: Toward unsupervised activity discovery using multi-dimensional motif detection in time series. In: *Proceedings of the 21st International Jont Conference on Artifical Intelligence*, Morgan Kaufmann Publishers Inc., 2009 (IJCAI'09 9), P. 1261–1266

[VDV20]   VU, H. ; DUONG, V. N. ; VU, V.: Exploiting the fine-grained similarity of a large-scale rice species using shape motif discovery. In: *2020 RIVF International Conference on Computing and Communication Technologies (RIVF)*, IEEE, 2020, P. 1–6

[VMP+18]   VARATHARAJAN, R. ; MANOGARAN, G. ; PRIYAN, M. K. ; SUNDARASEKAR, R.:   Wearable sensor devices for early detection of Alzheimer disease using dynamic time warping algorithm. In: *Cluster Computing* 21 (2018), Nr. 1, P. 681–690. – Springer

[VNK13]   VESPIER, U. ; NIJSSEN, S. ; KNOBBE, A.: Mining characteristic multi-scale motifs in sensor-based time series. In: *Proceedings of the 22nd ACM International Conference on Information & Knowledge Management*, ACM, 2013, P. 2393–2398

[WaM18]   WAGHMARE, P. A. ; MEGHA, J. V.: Efficient Pattern Recognition in Time Series Data. In: *2018 International Conference on Inventive Research in Computing Applications (ICIRCA)*, IEEE, 2018, P. 436–441

[WBK09]   WEICKERT, T. ; BENJAMINSEN, C. ; KIENCKE, U.:   Analytic Wavelet Packets-Combining the Dual-Tree Approach with wavelet Packets for Signal Analysis and Filtering. In: *IEEE Transactions on Signal Processing* 57 (2009), Nr. 2, P. 493–502. – IEEE

[Wei94]   WEI, W. W.-S.: *Time series analysis*. Addison-Wesley publ Read-
               ing, 1994. – ISBN 0201159112

[WFBS14]   WELLER-FAHY, D. J. ; BORGHETTI, B. J. ; SODEMANN, A. A.:
               A survey of distance and similarity measures used within network
               intrusion anomaly detection. In: *IEEE Communications Surveys &
               Tutorials* 17 (2014), Nr. 1, P. 70–91. – IEEE

[WFH11]   WITTEN, I. H. ; FRANK, E. ; HALL, M. A.: *Data Mining: Practical
               Machine Learning Tools and Techniques*. 3rd ed. Elsevier Morgan
               Kaufmann, 2011. – ISBN 9780123748560

[WGG+15]   WANG, R. ; GAO, J. ; GAO, Z. ; GAO, X. ; JIANG, H.: Hilbert-
               Huang transform based pseudo-periodic feature extraction of nonlin-
               ear time series. In: *Seventh International Conference on Measuring
               Technology and Mechatronics Automation*, IEEE, 2015, P. 532–537

[Whi51]   WHITTLE, P.: *Hypothesis testing in time series analysis*. 4th ed.
               Almqvist & Wiksells, 1951. – ISBN 9780598919823

[Wic96]   WICKERHAUSER, M. V.: *Adaptive Wavelet Analysis: From Theory
               to Software*. 1996. – ISBN 978–1568810416

[Wil17]   WILSON, S. J.: Data representation for time series data mining:
               time domain approaches. In: *Wiley Interdisciplinary Reviews: Com-
               putational Statistics* 9 (2017), Nr. 1, P. e1392. – Wiley Online Li-
               brary

[WLM+20]   WANG, F. ; LI, M. ; MEI, Y. ; LI, W.: Time series data mining:
               A case study with big data analytics approach. In: *IEEE Access* 8
               (2020), P. 14322–14328. – IEEE

[WMD+13]   WANG, X. ; MUEEN, A. ; DING, H. ; TRAJCEVSKI, G. ; SCHEUER-
               MANN, P. ; KEOGH, E.: Experimental comparison of representa-
               tion methods and distance measures for time series data. In: *Data
               Mining and Knowledge Discovery* 26 (2013), Nr. 2, P. 275–309. –
               Springer

[Wol88]   WOLBERG, G.: Geometric transformation techniques for digital
               images: a survey. In: *Department of Computer Science Columbia
               University* (1988)

[Woo96]   WOOD, J.: Invariant pattern recognition: a review. In: *Pattern
               recognition* 29 (1996), Nr. 1, P. 1–17. – Elsevier

[WRH18]   WIJAYANTO, I. ; RIZAL, A. ; HADIYOSO, S.: Multilevel Wavelet
               Packet Entropy and Support Vector Machine for Epileptic EEG
               Classification. In: *4th International Conference on Science and
               Technology*, IEEE, 2018, P. 1–6

[XKW+07]   XI, X. ; KEOGH, E. ; WEI, L. ; MAFRA-NETO, A.: Finding mo-
            tifs in a database of shapes. In: *Proceedings of the 2007 SIAM
            international conference on data mining*, SIAM, 2007, P. 249–260

[YCZ+15]   YAN, J. ; CHO, M. ; ZHA, H. ; YANG, X. ; CHU, S. M.: Multi-graph
            matching via affinity optimization with graduated consistency reg-
            ularization. In: *IEEE transactions on pattern analysis and machine
            intelligence* 38 (2015), Nr. 6, P. 1228–1242. – IEEE

[YeK09]   YE, L. ; KEOGH, E.: Time series shapelets: a new primitive for data
            mining. In: *Proceedings of the 15th ACM SIGKDD International
            Conference on Knowledge Discovery and Data Mining*, ACM, 2009,
            P. 947–956

[YKA+19]   YETIS, H. ; KARAKOSE, M. ; AYDIN, I. ; AKIN, E.: Bearing Fault
            Diagnosis in Traction Motor Using the Features Extracted from Fil-
            tered Signals. In: *2019 International Artificial Intelligence and Data
            Processing Symposium (IDAP)* IEEE, 2019, P. 1–4

[YKM+07]   YANKOV, D. ; KEOGH, E. ; MEDINA, J. ; CHIU, B. ; ZORDAN, V.:
            Detecting time series motifs under uniform scaling. In: *Proceedings
            of the 13th ACM International Conference on Knowledge Discovery
            and Data Mining*, ACM, 2007, P. 844–853

[YoC15]   YOSHIDA, S. ; CHAKRABORTY, B.: A comparative study of simi-
            larity measures for time series classification. In: *JSAI International
            Symposium on Artificial Intelligence*, Springer, 2015, P. 397–408

[YSR+13]   YINGCHAREONTHAWORNCHAI, S. ; SIVARAKS, H. ; RAKTHAN-
            MANON, T. ; RATANAMAHATANA, C. A.: Efficient proper length
            time series motif discovery. In: *13th International Conference on
            Data Mining*, IEEE, 2013, P. 1265–1270

[YTL+16]   YI, K. M. ; TRULLS, E. ; LEPETIT, V. ; FUA, P.: Lift: Learned
            invariant feature transform. In: *European Conference on Computer
            Vision*, Springer, 2016, P. 467–483

[Yu12]   YU, R.: Shift-Variance Analysis of Generalized Sampling Processes.
            In: *IEEE Transactions on Signal Processing* 60 (2012), Nr. 6, P.
            2840–2850. – IEEE

[Yu16]   YU, S.-Z.: *Hidden Semi-Markov Models: Theory, Algorithms and
            Applications.* 2016. – ISBN 978 0-12-802767-7

[YuO05]   YU, R. ; OZKARAMANLI, H.: Hilbert transform pairs of orthogo-
            nal wavelet bases: Necessary and sufficient conditions. In: *IEEE
            Transactions on Signal Processing* 53 (2005), Nr. 12, P. 4723–4725.
            – IEEE

[YWZ+19]   YIN, J. ; WANG, R. ; ZHENG, H. ; YANG, Y. ; LI, Y. ; XU, M.:
            A New Time Series Similarity Measurement Method Based on the
            Morphological Pattern and Symbolic Aggregate Approximation. In:
            IEEE Access 7 (2019), P. 109751–109762. – IEEE

[ZaY17]   ZAN, C. T. ; YAMANA, H.:   A Variable-length Motifs Discovery
            Method in Time Series Using Hybrid Approach. In: *Proceedings of
            the 19th International Conference on Information Integration and
            Web-based Applications & Services*, ACM, 2017 (iiWAS '17), P. 49–
            57

[Zha11]   ZHANG, X.: Design of Q-shift filters with improved vanishing mo-
            ments for DTCWT. In: *18th IEEE International Conference on
            Image Processing*, IEEE, 2011, P. 253–256

[ZhG16]   ZHONG, J. ; GAN, Y.: Detection of copy–move forgery using discrete
            analytical Fourier–Mellin transform. In: *Nonlinear Dynamics* 84
            (2016), Nr. 1, P. 189–202

[ZLG17]   ZHANG, L. ; LIU, P. ; GULLA, J. A.: A Neural Time Series Forecast-
            ing Model for User Interests Prediction On Twitter. In: *Proceedings
            of the 25th Conference on User Modeling, Adaptation and Person-
            alization*, ACM, 2017 (UMAP '17), P. 397–398

[ZLV16]   ZHANG, G. ; LILLY, M. J. ; VELA, P. A.: Learning binary features
            online from motion dynamics for incremental loop-closure detec-
            tion and place recognition. In: *IEEE International Conference on
            Robotics and Automation*, IEEE, 2016, P. 765–772

[ZWS09]   ZHOU, G. ; WANG, W. ; SUN, S.: Phase Properties of Complex
            Wavelet Coefficients. In: *2009 Fourth International Conference on
            Innovative Computing, Information and Control (ICICIC)*, IEEE,
            2009, P. 1373–1376

[ZYZ+18]   ZHU, Y. ; YEH, C. M. ; ZIMMERMAN, Z. ; KAMGAR, K. ; KEOGH,
            E.: Matrix Profile XI: SCRIMP++: Time Series Motif Discovery
            at Interactive Speeds. In: *IEEE International Conference on Data
            Mining*, 2018, P. 837–846. – IEEE

[ZYZ+20]   ZHU, Y. ; YEH, C. M. ; ZIMMERMAN, Z. ; KEOGH, E.:   Matrix
            Profile XVII: Indexing the Matrix Profile to Allow Arbitrary Range
            Queries. In: *2020 IEEE 36th International Conference on Data
            Engineering (ICDE)*, IEEE, 2020, P. 1846–1849

List of Tables

© The Editor(s) (if applicable) and The Author(s), under exclusive license to
Springer-Verlag GmbH, DE, part of Springer Nature 2022
S. Deppe, *Discovery of Ill-Known Motifs in Time Series Data*, Technologien
für die intelligente Automation 15, https://doi.org/10.1007/978-3-662-64215-3

List of Figures

© The Editor(s) (if applicable) and The Author(s), under exclusive license to
Springer-Verlag GmbH, DE, part of Springer Nature 2022
S. Deppe, *Discovery of Ill-Known Motifs in Time Series Data*, Technologien
für die intelligente Automation 15, https://doi.org/10.1007/978-3-662-64215-3

Printed in the United States
by Baker & Taylor Publisher Services